Monograph Series   Volume 2

MIDDLE MATH

Improving the Undergraduate Preparation of Teachers of Middle Grades Mathematics

Edited by

Mary B. Eron
Sidney L. Rachlin
East Carolina University

Monograph Series Editor

Denisse R. Thompson
University of South Florida

Association of
Mathematics Teacher Educators

Published by the Association of Mathematics Teacher Educators, San Diego State University, c/o Center for Research in Mathematics and Science Education, 6475 Alvarado Road, Suite 206, San Diego, CA 92120

www.amte.net

Support for the MIDDLE MATH Project was provided by the National Science Foundation under Grant No. DUE 9455152, the North Carolina Statewide Systemic Initiative, and East Carolina University. The views expressed or implied in this publication, unless otherwise noted, should not be interpreted as official positions of the Association of Mathematics Teacher Educators or any of the agencies that provided financial support for its production.

Library of Congress Cataloging-in-Publication Data

MIDDLE MATH: improving the undergraduate preparation of teachers of middle grades mathematics / edited by Mary B. Eron, Sidney L. Rachlin.

    p. cm. — (Monograph series ; v. 2)
Includes bibliographical references
ISBN: 978-1-62396-943-1
    1. Mathematics—Study and teaching—United States. 2. Mathematics teachers—Training of—United States.
I. Eron, Mary B., 1957-2005. II. Rachlin, Sid, 1947-. III. Series: Monograph series (Association of Mathematics Teacher Educators); v.2.
    QA11.2E67 2005
    510.71'2—dc22

Library of Congress Catalog Control Number: 2006000298
International Standard Book Number: 1-932793-03-8

MIDDLE
MATH
PROJECT

Table of Contents

Part II: Case Studies

Part III: Issue Papers

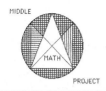

Foreword

The following monograph represents the work of many mathematics teacher educators who explored the content knowledge and pedagogical knowledge that make up the middle grades learning experience. The middle grades remains a unique period of time in students' development and as such provides both challenges and promising opportunities for those who prepare teachers of middle grades mathematics.

This work is the final product of an exciting NSF supported endeavor that gathered leaders in the field and explored curriculum, case studies of program models at several institutions, as well as issue papers on such key topics as assessment, technology, and preparing culturally responsive teachers. Although this work was completed in 1998, there are many key components of this document that are just as useful and meaningful today as they were then. AMTE has decided to publish *Middle Math: Improving the Undergraduate Preparation of Teachers of Middle Grades Mathematics* as a resource to AMTE members, especially those who are revising and refining their middle school programs. We hope this monograph will stimulate discussion and bring attention to this critical period of schooling.

We are grateful for the work of Sid Rachlin at East Carolina University and Denisse Thompson at the University of South Florida who have brought this important monograph to fruition. AMTE hopes that this document will be useful to the members as key readings for prospective teachers as well as a catalyst for conversations with university faculty and school partners.

> *Karen S. Karp*
> *AMTE President 2003–04*

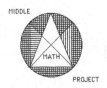

MIDDLE

MATH

PROJECT

Acknowledgements

We would like to thank the many people who helped make the MIDDLE MATH Project a success and, in particular, those whose time and effort went into the production of this monograph.

MIDDLE MATH Advisory Board
Sunday Ajose, East Carolina University
Robert L. Bernhardt, East Carolina University
R. Michael Hoekstra, Centers for Disease Control, Atlanta, Georgia
Ann Hutchens, North Carolina Department of Public Instruction,
     Northwest Technical Assistance Center, Wilkesboro
John E. Owens, Savannah High School, Savannah, Georgia
Ron Preston, East Carolina University
Mary Kim Pritchard, University of North Carolina at Charlotte
Rose Sinicrope, East Carolina University
Michael Spurr, East Carolina University

MIDDLE MATH Editorial & Production Staff
Kathy Stanley, Project Manager
Jennifer Jacobs, Administrative Assistant
JoAnn Harrington, Graduate Assistant
Brett Hursey, Graduate Assistant and English Editor
Caren Penny, Graduate Assistant and English Editor
Lara Smith, Graduate Assistant
Stephanie Woodley, Graduate Assistant
Gemma Foust, Undergraduate Assistant
Amanda Stanley, Undergraduate Assistant
Cheryl Johnson, Teacher-In-Residence

We would also like to gratefully acknowledge the funding we received from the National Science Foundation, the North Carolina Statewide Systemic Initiative, and Texas Instruments Incorporated without which our project could not have succeeded.

Finally, we would like to thank Denisse Thompson, Ed McClintock, and Zhonghong Jiang for keeping the publication alive and Karen Karp, Skip Fennell, and the 2003 AMTE Board of Directors for giving us the opportunity to share it with the AMTE membership.

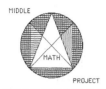

Introduction

Background

In the mid-1980's, programs were implemented to prepare undergraduates to be certified as middle school mathematics teachers; these programs borrowed courses from already existing elementary or secondary programs. The result—an unfortunate hodgepodge of preparation—did not specifically address the unique needs of middle school teachers. From 1986 to 1991, the NSF funded nine projects to develop teacher preparation programs for middle school mathematics and science teachers. Those that involved mathematics include the *Middle Grades Teacher Preparation Project* at the University of Georgia, the *Middle School Math Program* at Illinois State University, *Preparing Pre-service Middle School Science and Math Teachers* at Lesley College, the *Development, Implementation, and Research for Educating Competent Teachers* (DIRECT) *Project* at Oklahoma State University, the *Middle School Math Project* at Portland State University, and the *Model Teacher Preparation Program* at SUNY-Potsdam.

A program for the preparation of teachers of middle grades mathematics should address the development of five types of knowledge: *Knowledge of Mathematics, Knowledge of the Teaching of Mathematics, Knowledge of the Learning of Mathematics, Knowledge of the Learner of Mathematics*, and *Knowledge of School Mathematics*.

Knowledge of Mathematics. The first issue in the design of a program of study for prospective middle grades mathematics teachers is to decide the precise mathematical content of the undergraduate curricula. In *A Call for Change*, the MAA recommends that "*special courses should be developed that provide the proper focus and breadth of experience for these teachers*" (Leitzel, 1991, p. 17). There is no clear consensus on what this curriculum should be. *On the Shoulders of Giants* offers themes that could be used to develop innovative course materials (Steen, 1990). A more traditional approach would be to start with one of the several lists of 'big ideas' in mathematics as a basis (Leitzel, 1994). A third approach would be to look at the changing nature of the middle grades curricula and design courses to provide a

foundation and an overview for this content; for example, an undergraduate class on probability and statistics could be designed which
would support the NCTM *Curriculum and Evaluation Standards*
(1989). These possibilities need to be developed into courses and
programs.

Knowledge of the Teaching of Mathematics. Attention needs to be
paid to the pedagogy of the college classroom as well, for many feel
that teachers teach as they were taught (National Research Council,
1989). Given the increasing acceptance of the constructivist theory of
learning, the college classroom must become a more active place. Preservice teachers must experience the climate that they will be expected
to create in their classroom. Teacher preparation programs must be
attentive to this need.

Knowledge of the Learning of Mathematics. In terms of the children
in the classroom, middle grades mathematics instruction is crucial in
determining attitudes about the value and the usefulness of the subject
and hence continued interest in studying mathematics (Leitzel, 1991;
NSF, 1993). It is vital to train teachers of mathematics who know the
wonder and beauty of the subject and can transmit it to their students.
Peter Hilton, speaking at an East Carolina University (ECU) seminar
series on improving the undergraduate preparation of teachers of
mathematics, commented that the particular topics in mathematics
learned by future teachers are not as important as their attitudes toward
mathematics. He spoke of conveying the beauty of mathematics and the
esthetic satisfaction of doing mathematics.

Knowledge of the Learner of Mathematics. Prospective teachers of
middle school mathematics must be students of the psychology of
learning and instruction of middle grades mathematics. During the
middle school years, students are taught the mathematics of rational
numbers and the operations of multiplication and division. Although it
would appear that rational numbers are simply a generalization of
whole numbers and that multiplication and division are just more
advanced concepts than addition and subtraction, the learning of these
concepts is much more complex than the mathematics of primary
grades. Psychologically, they require the difficult process of reconstructing mental structures. Unfortunately, most students progress
through the middle school grades without developing these mathematical concepts (Hiebert & Behr, 1988). In order to provide instruction
that will confront the student with the need to re-construct their concepts, teachers need to understand the developmental process for each

concept and must be able to assess the students' developmental stages in order to provide appropriate instruction (Fuson, 1988). Rather than waiting for students to make the transition from concrete to formal operational reasoning, many mathematical concepts can be taught and learned at a concrete level (Kloosterman & Gainey, 1993).

With the development of the Van Hiele levels of geometric understanding, middle school teachers have a framework by which necessary and appropriate geometry instruction can be selected, designed, and implemented (Geddes & Fortunato, 1993). Similarly, we now have a better understanding of how students develop algebraic concepts and which algebraic concepts can be effectively constructed by the middle school student (Wagner & Kieran, 1989). The research on the learning of mathematics should inform teachers as they make decisions in the classroom. Therefore, in the course of their undergraduate education, prospective middle school teachers should learn about this research and its use in the classroom.

Knowledge of School Mathematics. Middle level mathematics is important in the overall scheme of things, yet it has been criticized as a curriculum standing still. Very little new material has traditionally been covered in the course of grades five through eight (Flanders, 1987; NCTM, 1989). The NCTM *Curriculum and Evaluation Standards* attempted to address this problem by emphasizing the need to introduce pre-algebra and function concepts earlier in the curriculum.

It is important for university faculty to be aware of how the changes called for by NCTM have been interpreted and extended by these curriculum developers. University faculty need to know what their students will be expected to teach and how they will be expected to teach it. Teacher preparation programs should reflect these expectations. It is better to improve rather than repeat mathematics education (National Research Council, 1989).

A familiarity with the future middle school curriculum is, however, only a beginning. The next and most important step is to decide on the mathematical background desired for the teachers of this curriculum. Teachers need knowledge and understanding considerably deeper than what they are required to teach (Leitzel, 1991). They need to know how the mathematics builds on itself so they know what to emphasize in their classrooms (Hilton, 1994). The problem of identifying the nature of the mathematics needed by middle school teachers was the driving force behind NSF's initiative to fund Middle School Science and

Mathematics Teacher Preparation Projects (Stake et al., 1993). The insights gained from these experiences should be useful in designing new programs.

We can also learn from the collective lessons of past experience. In its design, the MIDDLE MATH project addresses two criticisms raised in the *Teacher Preparation Archives*. First, in implementing any reform movement it is important to debate the issues, not to accept change blindly, and to look at underlying assumptions. This is why we felt that our program needed a national scope. Also, since all change is local, global models need to be adapted to the situation in home institutions (National Research Council, 1989; Stake et al., 1993).

In addition to the content and organization of preparation programs, it is necessary to understand how the undergraduate student learns. Research on how adults learn mathematics is now evolving (Ferrini-Mundy, 1994; Selden & Selden, 1993). Because of the changes in school mathematics, students entering college in the future will be different. The effects of these changes on the entering freshman and the implications these changes have for the undergraduate mathematics curriculum are currently being studied (Leitzel, 1993). One important difference will be students' greater knowledge and experience with technology. It is important to design the college curriculum carefully to take advantage of this new knowledge (Leitzel, 1991).

There is compelling evidence that existing teacher education programs will not produce the teachers with the knowledge and understanding necessary to teach the mathematics envisioned by the mathematics education community (Brown & Borko, 1992) or to handle the challenge of a diverse, multi-cultural population at the difficult stage of maturing emotionally, physically, sexually, morally, socially, and cognitively. We are compelled to reflect, study, and change our teacher preparation programs.

MIDDLE MATH Beginnings

MIDDLE MATH (1995-98) was an Undergraduate Faculty Enhancement Project funded by the National Science Foundation (DUE 9455152), the North Carolina Statewide Systemic Initiative, East Carolina University, and Texas Instruments. The MIDDLE MATH Project was designed to focus attention on improving the undergraduate preparation of teachers of middle grades mathematics.

The project started when, in the Spring of 1993, the East Carolina University's Mathematics Education faculty set as a goal the redesign of their program for preparing middle grades teachers of mathematics. A group of faculty from the Mathematics and Mathematics Education areas formed a task force to develop a longitudinal plan for improving the middle grades program. The Math Department provided funds to support a seminar series on improving the undergraduate preparation of teachers of mathematics. In addition to presenting a colloquium, each consultant met for several hours with the department's middle grades task force. The areas discussed with the consultants included the mathematics content, pedagogy, and structure of an improved preparation program for prospective middle school teachers of mathematics.

Based on the consultants' responses, the task force decided it needed much more input before it could comfortably set the design of a new middle grades program. The MIDDLE MATH project was thus conceived. The project reflects the needs identified by the task force: the commitment from additional mathematics and mathematics education faculty to participate in and collaborate on the project; a better sense of the changing content and pedagogical knowledge (reflective of the calls for reform) required to teach middle grades curricula; an increased familiarity with the experiential knowledge accrued by others who have set down this path; and an awareness of how the growing body of research on the teaching and learning of middle grades and undergraduate mathematics might impact a teacher preparation program.

MIDDLE MATH Goals and Objectives

The Middle Math project:

1. Supports the design of new models for the mathematics component of middle grades teacher preparation programs by providing an opportunity for university mathematicians and mathematics educators to:

    *   learn about new and innovative middle school curricula that model and extend the NCTM *Standards*,

    *   discuss the influences of middle grades curricula on teacher preparation programs with particular attention to content, pedagogy, technology, and management of a diverse student population,

    *   analyze previous programs that addressed the issues of training middle school teachers,

- define the mathematics content preservice teachers should learn and how the content should be structured and taught,
- explore how changes advocated by the NCTM and the MAA should effect teacher preparation programs,
- review the research on teaching and learning of middle school mathematics and identify the resources that prospective teachers will need to continue their education as practicing teachers, and
- reflect on the changes in the undergraduate learner as a result of changes in the curriculum in the schools and anticipate the effect of the changes on college instruction.

2. Facilitates the development of working relationships between colleagues to support individual and team efforts to make program changes.

3. Encourages the implementation of these programs by:
   - providing support networks;
   - informing participants about funding sources for curriculum development;
   - sharing the curriculum implications of research on the teaching and learning of undergraduate mathematics.

The project was divided into five phases: recruitment and preparation, first summer conference, academic-year curriculum development, second summer conference, and dissemination.

Recruitment and Preparation

With the addition of new members, the task force became the MIDDLE MATH Advisory Board. Comprising the board are three mathematicians, a statistician, and five mathematics educators on the faculty of East Carolina University. During the spring semester of 1995, the board met twice a week, exploring a variety of ways to facilitate revision of the undergraduate preparation program for teachers of middle grades mathematics. They read and discussed documents from the various mathematical organizations calling for reform, reviewed the publications proposed as background reading for conference participants, and met with middle grades mathematics teachers to discuss their recommendations for changing undergraduate teacher preparation. They generated discussion questions and wrote reflections on the discussions they had. This shared interaction helped the advisory board design "homework" for MIDDLE MATH participants and refine the first conference agenda.

The recruitment phase was important for raising a national conscious-ness of the problem, as well as increasing the likelihood of participation of women, underrepresented minorities, and persons with disabilities. Notices were sent out. Nationally (that is, excluding North Carolina), the project received approximately one hundred applications from sixty-two different institutions. Of them, twenty-four people were accepted as paid participants (six others were invited and paid their own expenses). The chance to participate, therefore, was roughly one in four.

A natural diversity of perspectives for program development was encouraged by giving preference to collaborative teams (for example, mathematicians and mathematics educators from the same institution or faculty from a university and feeder two-year college). In order to maximize the project's potential impact, a high priority was put on local collaborations—the momentum for change is greatly enhanced when several of those involved understand and appreciate the unique level of peculiarities at a given college or university. (Recognizing this, the North Carolina Statewide Systemic Initiative supported the partici-pation of faculty from universities, colleges and community colleges in North Carolina).

Out of the thirty-two applications from North Carolina, twenty-nine were accepted as participants representing fifteen different institutions. Because of the large number of applications, only sites with two or more applicants from outside of North Carolina were included. Selec-tions were also based on the quality of the responses to the open-ended questions on the application.

The project's fifty-nine participants included representatives of four-year institutions empowered by their states to prepare teachers of middle level mathematics for certification or endorsement, and four participants from two-year feeder colleges (with mathematics courses that transfer into the middle grades level mathematics programs of the four-year institutions). Participants included research, teaching, and administrative faculty. The mathematics faculty included both pure and applied mathematicians as well as statisticians. The mathematics education faculty included faculty actively pursuing research on the teaching of mathematics, the learning of mathematics, and/or the design of curricula to support the teaching and learning of mathematics.

Once selected, participants prepared for the first summer conference by

1. reviewing the National Council of Teachers of Mathematics
    (NCTM) *Curriculum and Evaluation Standards for School Mathe-
    matics* for grades 5-8 and the Mathematical Association of
    America (MAA) recommendations for changing the nature of the
    mathematical preparation of the middle grades teacher, *A Call for
    Change,*

2. reviewing and reflecting on *Teacher Preparation Archives: Case
    Studies of NSF-Funded Middle School Science and Mathematics
    Teacher Preparation Projects, 1986-91* which offered some insight
    into what nationally recognized work has already taken place in
    teacher preparation,

3. considering the nature of college mathematics that the middle
    grades teacher should experience as reflected in the text *On the
    Shoulders of Giants*, and

4. interviewing a middle grades teacher to ensure that teachers' input
    and viewpoints would be heard and incorporated into the process
    even though middle grades teachers would not be present at the
    conferences.

The five bi-weekly "homework" assignments the project sent to participants included suggested readings, a set of focus questions, and a
journal assignment. This shared knowledge base would be used as a
foundation for discussions at the MIDDLE MATH conferences.

Participants also gathered information on their state's requirements and
guidelines for the preparation of middle grades mathematics teachers,
their state's current curriculum for middle grades students, and their
own current programs and courses for preparing middle grades teachers. This information was shared at the first summer conference so that
participants could compare their current middle grades teacher preparation program with other programs and program models.

The First Summer Conference

The MIDDLE MATH First Summer Conference was held from August
2–6, 1995 at East Carolina University in Greenville, North Carolina.
The conference provided an opportunity for participants to look back at
their own programs after reviewing five NSF-funded Middle School
Mathematics Teacher Preparation Projects (1986-1991), and to look
ahead to the changing needs of future middle-level mathematics teach-

ers and the changing nature of the mathematical background, modes of instruction, and assessment methods of the high school graduate.

A Look Back: Voices of Experience

The successes and failures of the NSF-funded Middle School Mathematics Teacher Preparation Projects should inform any new attempt at designing teacher preparation programs. Therefore, the Middle Math Project invited representatives from these projects to speak at the first conference. Jane Swafford from Illinois State University, James Choike from Oklahoma State University and Michael Shaughnessy from Portland State University were able to attend, sharing activities and speaking about their projects. They gave brief overviews of their programs and talked about what had gone well, what had not, and the changes they made in their programs and courses (from those proposed in the original NSF grant) to keep a successful and viable middle grades mathematics program thriving at their institutions. Each offered experientially-based suggestions on how to begin to design or redesign such programs at other institutions.

Middle School Math Program: Illinois State University
Jane Swafford

The *Middle School Math Program* has evolved since its NSF-project funding. Consisting of 30 semester hours of mathematics (including 18 hours mandated by the state that every elementary education major must take in either reading, writing or arithmetic), the program requires courses in problem solving, probability and statistics, and calculus as well as a capstone modeling-course—all specifically designed for the K-8 teacher.

Swafford discussed a variety of assessment tools used throughout the program at Illinois State, describing, in particular, many of the alternative assessment ideas she has experimented with in her elective course on Modern Algebra. Three of these stand out: first, using reflective journals as a vehicle for students to talk to the teacher about the troubles and successes they experience in a course. Students seemed to find writing much easier than coming up to the teacher's office to talk about the course. Second, Swafford advocated the use of portfolios. For her Modern Algebra portfolio, she had students submit four pieces, one for each of the NCTM *Standards* processes: problem solving, communication, reasoning, and mathematical connections. For the connection piece, students had to write an essay about the connection between modern algebra and K-8 curriculum. "While they would not believe me [at first]," Swafford comments, "what they find is amazing to them. When they write this essay, and many of them do a detailed analysis of the K-8 textbooks, they come back saying 'It's all the way through this'." The students end up with a real understanding of why they have spent time learning Modern Algebra. Perhaps, they see how the mathematics builds on itself so they know what to emphasize in their classrooms (Hilton, 1994). The value of reflection is in helping students make connections between what they observe in classrooms, what they read, and what they believe.

Swafford's advice to those redesigning their middle grades program: first, be very cognizant of state requirements in designing a program. Once the program is established, she recommends recruiting heavily — use posters, visit prerequisite classes, put advertisements in the student newspaper, hold informational forums in the evenings, and send letters to students with high grade point averages and undeclared majors. Their selling point: "If you want a good job, have a concentration in mathematics." Her one caution was to rethink the pedagogy in mathematics courses. Each of the courses in the program was developed with a mathematician and a math educator. She warned against mathematicians going it alone. Involve math educators in the development of each course. The partnership between educators with these two backgrounds is necessary and powerful.

DIRECT: Oklahoma State University
James Choike

Jim Choike, of the DIRECT project at Oklahoma State, espoused a different philosophy, feeling that you should not design mathematics courses especially for middle grades teachers. Rather, you should design good (interactive and problem-centered) mathematics courses that could be taken by any math major. This philosophy grew out of the NSF grant to Oklahoma State to design both a science and mathematics program for middle grades teachers. The science part of the program, with its specialized courses that no one wanted to teach, no longer exists, while the mathematics part is thriving. Also contributing to his philosophy is his firm belief that those preparing to teach mathematics need to see and learn mathematical content in the same way that people are going to do mathematics. The strong and rigorous mathematical content of the DIRECT program reflects these beliefs.

Choike spoke about the difficulties of creating a new program, mentioning the various stakeholders, all with criteria to be satisfied: individual faculty, departments (mathematics and curriculum and instruction), college (Arts and Sciences and Education), and, of course, state Departments of Education. He told stories of faculty who would not participate due to a lack of professional incentive in departments where only research is valued, or fearful of losing their pet course. He emphasized the need for extensive pre-planning, unilateral faculty and administrative involvement, and especially, building consensus.

Choike echoed Swafford's belief regarding the importance of giving prospective middle grades teachers a consistent model for teaching. He was greatly troubled that while all the new mathematics courses in the program embraced such inquiry-based instruction (that is, learning-by-discovery teaching methods), the majority of courses taken by the students in the program were lecture-based. He felt these frustrated students, creating a concern needing to be addressed in DIRECT, as well as any middle grades mathematics program. His last bit of advice was "when you are designing programs, ownership is a key word." The more ownership you have invested in a program, the more you will invest in seeing it continue.

Middle School Math Project: Portland State University
Michael Shaughnessy

Because of the characteristics of Portland State, the middle grades program developed there differed from the others described at the conference in three significant ways. First, all of the courses emphasized visual thinking as an important tool for doing mathematics. The Math Learning Center housed at Portland State, a not-for-profit math education company, has developed K-8 curriculum materials emphasizing visual thinking and has established extensive professional development programs. Second, the people who worked on the *Middle School Math Project* were also working on curriculum development at the Center at the same time; Shaughnessy observed that this led to "very, very dramatic" cross connectors between these two projects. As with the other projects creating teacher preparation programs, discussion within a community of scholars was key to successfully developing the program. Third, because of the in-service programs at the Center and the urban location of the university, most of the students in the middle school math program are already teachers working part-time on a masters degree, obtaining a middle school certificate or an endorsement in math. Although all of the courses in the program are co-

listed as undergraduate courses, most of the students are graduate
students.

Prior to this program at Portland State, middle school teachers were
taking courses with secondary teachers, finding them very inappropriate for their needs. All eight courses in the program have "for Middle
School Teachers" in their title (e.g., Concepts of Calculus for Middle
School Teachers). They include Computing in Mathematics, Experimental Probability and Statistics, Problem Solving, Geometry, Arithmetic and Algebraic Structures, Historical Topics, and a methods
course. The courses, all taught in the mathematics department, run
cyclically, and students can start the program with any one of them.

Program courses include no lectures, centering instead on small-group
problem solving. Visual models are heavily used. Shaughnessy feels so
strongly about the need to emphasize visual thinking that he wants
visual models included in the list of multiple representations used in
mathematics (along with tables, graphs and symbols). In addition to
students "discussing and listening to how others think, and sharing
different approaches to problems" in class, the program strongly
emphasizes a writing component. Students are encouraged to write
about how they got to solutions of problems or how they thought about
a problem. Shaughnessy describes passing a folder of work back and
forth between student and teacher. The students write, the teacher
comments on their writing, and then the students respond. To be able to
engage in a reflective analysis of one's thinking processes is one of the
program goals. In all of these courses, students are asked to write
comments on their own growth and development several times
throughout the course. They talk about things they are learning, things
they are stuck on, and things they still need to do.

Conclusions from *A Look Back*

In comparing the preparation programs for teachers of middle grades
mathematics developed at these three universities, one is struck by
some key similarities. Most obvious is the need to change from lecturing to a more interactive and problem-centered mode. All three speakers emphasized the crucial nature of a cooperative learning environment based on a constructivist philosophy. They seek to model the type
of teaching they expect their students to use in the middle grades
classroom. Secondly, collaboration and cooperation between mathematicians and mathematics educators in creating courses enriches the
courses and programs in ways far greater than may be expected. Third,

because programs differ primarily based on the individual needs of the schools, departments or personnel involved, each program must emphasize the strengths of the university where it is developed.

Conference participants benefited from interactions with the speakers in several ways. Some found confirmation of ideas they already held, glad to find their current or proposed program in line with the ideas of others. Some found new ideas they had not tried before, some precisely what they were looking for in their program (e.g., the calculus materials developed at Portland State were mentioned several times). Resources were freely shared, from course guides to modeling problems. Discussions of anticipated difficulties in designing a program and ways to cope with them were sought out. The speakers interacted with the participants throughout the conference.

Striking were the differing views of mathematics which we hold. One speaker emphasized the visual; another the need for rigor; another the creative nature of mathematics. One speaker believed the preservice teacher's attitude toward mathematics is the most important thing to be molded in a preparation program. Another commented that to be a facilitator in a mathematical discussion or to teach mathematics on an as-needed basis, the preservice teacher needs a depth of mathematical knowledge; yet another preparation program is remarkable for its breadth of mathematics. To design an effective teacher preparation program, it seems we need to air our views of mathematics and, where possible, come to a consensus upon what to base the program. The MIDDLE MATH Project offered everyone involved an opportunity to engage in this debate.

NSF-funded Middle Grades Mathematics Curricula

In addition to benefiting from these speakers' insights, conference participants spent a significant portion of their time actively introduced to the five NSF-funded Middle Grades Mathematics Curricula then under development (*Connected Mathematics, Mathematics in Context: A Connected Curriculum for Grades 5-8, Middle-School Mathematics through Applications Project, Seeing and Thinking Mathematically Project,* and *Six Through Eight Mathematics*). With the NCTM *Standards* as a starting point, they offer five different possibilities for the middle school curriculum in the twenty-first century.

Prior to the conference, each NSF-funded Middle Grades Curriculum Project wrote a paper answering "What are the characteristics of a preservice program that would prepare teachers to implement your curriculum effectively?" Their responses were to center around the themes of *Knowledge of Mathematics, Knowledge of School Mathematics, Knowledge of the Teaching of Mathematics, Knowledge of the Learning of Mathematics,* and *Knowledge of the Learner of Mathematics*. These themes set by the advisory board recurred throughout the first conference, curriculum development and subsequent presentations on it at the second conference, and in the projects' evaluation materials. Establishing these five areas as a way to look at the needs of preservice teachers enabled all involved to delve deeper into how to structure a program to meet these needs. Each project's written responses were given to the conference participants at the end of its workshop. Details of some of the unique features of these curricula and the implications they have for preparing teachers for their implementation are provided later as Part I of this monograph.

On the third day of the conference, participants divided into working groups of about five persons each, discussing for two hours issues raised by their interactions with the new middle school curricula. The morning concluded with a panel discussion. The panel, comprised of the representatives from the five middle grades projects, reacted to questions on the implications of their curricula on changing university programs for preparing middle grades teachers. Discussed was the need for teachers to have a deep mathematical understanding in order to make connections between mathematical concepts and the activities and problems in the curricula. Also discussed were the implications of progressive mastery, alternative assessment, the need to understand student thinking, and the need to be more reflective as a teacher. Lesson plans need more flexibility to deal with starting and stopping breaks, and the whole learning environment in which preservice teachers are taught mathematics needs to be more active. Through the panel discussion, participants heard and reacted to developers' perspectives on how curriculum change should affect teacher preparation.

In the afternoon, participants returned to their working groups. This time the goal was to arrive at a plan for what they would attempt to do to improve the middle school teacher preparation at their home institutions. The representatives from the NSF-funded Middle School Mathematics Teacher Preparation Projects (1986-1991) participated in these working groups. Discussion began by brainstorming a wish list for an ideal middle grades mathematics teacher preparation program, initially assuming no institutional or collegial barriers. Participants discussed the input they received from their interviews with middle grades teachers, and then shared those aspects of their existing programs that they felt were already in line with their changing goals and directions. They also suggested new courses or modifications in existing courses that they would like to see. By the final morning of the conference, the working groups had re-formed around areas of common interest, including courses to be worked on (e.g., Algebra and Modeling) or networking options (e.g., North Carolina schools). Participants soon designed skeletons of new course syllabi for the courses they were proposing to develop during the following academic year and outlined a tentative model of what the middle grades teacher preparation program should look like at their institution.

After a report to the large group from each of the working groups, the conference closed with a panel presentation offering three reflections on the conference proceedings. Robert Bernhardt spoke from the perception of a mathematician, Sunday Ajose from that of a mathematics educator, and Ann Hutchens as a middle grades specialist. These members of the project's advisory board reminded participants that some of the responsibility for the success of reform must be placed on students as well as educators, that we have yet to develop a model for training facilitators of learning, and that we must take the needs and characteristics of the young adolescent learner into account. They provided closure by looking back at their own experiences during the conference and looking ahead to the goals and directions for the remainder of the project.

Academic-Year Curriculum Development

During the academic-year (1995-96) of curriculum development, each MIDDLE MATH participant was expected to revise or develop both a model for the middle grades teacher preparation program at their institution and at least one course to be used in the preparation of middle grades mathematics teachers.

The AMTE bulletin board provided a vehicle to allow MIDDLE MATH participants to use electronic mail to maintain *across the hall* collegial relationships in which they could raise issues, ask specific nitty-gritty questions, make comments, and share working papers.

In addition to the AMTE bulletin board keeping participants loosely connected and informed as to what others were doing, participants were sent a newsletter summary of how some were progressing with the tasks they undertook as participants in the MIDDLE MATH Project.

This information came from the status report participants were asked to file, which answered the following questions:

1. How have you revised or developed a middle grades program for the preparation of mathematics teachers at your institution?

2. What course(s) are you modifying or developing and testing to fit into this program? How are you progressing with the course(s)?

Participants were expected to make presentations on their efforts to improve their middle grades teacher preparation programs at the second MIDDLE MATH conference.

Second Summer Conference

The MIDDLE MATH Second Summer Conference was held from June 7–9, 1996 at the Best Western Hotel in the Research Triangle Park and the Friday Center of the University of North Carolina at Chapel Hill. The second conference served as a working-meeting in which participants shared with each other the insights they had gained over the last year in trying to improve their middle grades math program for pre-service teachers. Therefore the same participants were invited back for the second conference. A total of 49 participants attended, 42 of whom had attended the First Summer Conference; the others were either replacements for participants who could not return or guests who attended at their own expense.

The second conference was designed to provide project participants with a catalyst for change. Participants were asked to prepare drafts of their proposed program models, course syllabi, and any related publications on which they were working. These were collected and distributed to the participants prior to the conference.

Participants were also asked to present information regarding the courses and programs they created and tested. Fourteen posters were presented at the conference. Poster sessions had a poster and handouts set up for two hours on the first day and a half of the conference. Groups of listeners would rotate through the posters every fifteen minutes so that the presenters could speak to each group about what they had done. Although the amount of time allotted to any one poster was short, participants got enough information to know who they wanted to talk to in greater detail later and received valuable handouts as well as some good project and activity ideas.

As an alternative to poster sessions, participants had two other options on how to present this material. The second option was a formal presentation of about 30 minutes. They could speak on the course(s) they had modified, the structure of their middle grades program, or a specific topic in teacher preparation that fit with what they had been working on. The third option was to facilitate or make a significant contribution to a "talk about" session on a variety of issues, including some that the participants suggested. Participants participated in three out of a possible nine of these "talk-about" discussion groups whose topics were: *Intra-University Collaboration and Other Political Issues*; *Changing Pedagogy*; *Data Analysis, Statistics and Probability*; *Field Experiences*; *Number/Algebra/Pre-Calculus/Calculus*; *Math Modeling/Discrete Math/Geometry*; *Alternative Forms of Student Assessment*; *Technology*; or *Math Methods Courses*.

In addition to the presentations by conference participants, two keynote speakers shared their insights—Lee Zia from the National Science Foundation, Division of Undergraduate Education, and Joan Ferrini-Mundy, Director of the Mathematical Sciences Education Board, National Research Council. Zia addressed the issue of funding; he talked about the nature of the NSF, how to write a grant proposal, gave examples of currently funded grants, and cited the existing sources of funding supporting the development of new curricula. This information gave participants insight into opportunities that future NSF funding may provide for improving the undergraduate preparation of teachers of middle grades mathematics.

Ferrini-Mundy spoke as a member of the faculty of the University of New Hampshire on *The Preparation of Teachers of Mathematics: Considerations and Challenges*, which is the title of a letter report from MSEB that she helped produce. First, she questioned the subject matter of teacher preparation and how we might view the intersection of mathematics and pedagogy. She suggested that we are engaged in the study of mathematics teaching problems. Teacher educators could use proxies for middle grades classrooms, such as video tapes, case studies, middle grades teachers' journals with classroom episodes, new middle grades curriculum materials, or actual middle grades classrooms with a change of focus to study with their preservice teachers how to manage the dilemmas that arise when teaching mathematics, to study mathematics in the context of teaching problems. She went on to challenge the assumptions on which we have based reform and offer questions that need to be further researched: Do we know if teachers teach as they were taught? How are real world experiences in industry or scientific studies beneficial for prospective teachers' future teaching? Given all of the different prevailing views about mathematics education, how do we position teachers to effectively defend and articulate to parents what they're trying to do? The issues and challenges she highlighted provided considerable discussion and food for thought.

Working groups were established to help participants review and refine their products. With the information provided them on sources for externally funding the development of their proposed curricula, and ways to use research on the learning of undergraduate mathematics to support and inform their curriculum development efforts, participants were encouraged to continue their reform efforts. Some participants were additionally provided travel support to share their efforts with others at professional meetings.

The conference closed with reports and summaries of the talk-about session and a small group discussion of the content and format of the MIDDLE MATH monograph. The monograph is divided into three parts. Part I includes abstracts prepared by the five NSF-funded Middle Grades Curriculum Programs and a statement concerning the changes that will need to be made in our undergraduate programs to prepare teachers to implement the new curricula.

Rather than summarize the participants' sessions for the monograph, we gave participants the opportunity to write a case study for their institution in which they highlighted some of the ways they were attempting to improve the preparation of teachers of middle grades mathematics. Participants submitted proposals and thirteen *cases* were selected based on the size and type of program involved. Part II consists of the case studies from these MIDDLE MATH colleges and universities. Each case study provides some detail on the nature of the institution, the program they have for preparing middle grades mathematics teachers, the accomplishments they have made in improving their program, and the challenges and issues that remained at that time. A *Case Study Matrix* is provided to help the reader locate which cases address specific topics of interest. Collectively, the case studies provide a series of snap shots of the initial efforts made by colleges and universities to change the way they prepare teachers for the middle grades mathematics classroom. While the universities have no doubt moved on from these initial efforts, the cases can be used to measure your own university's initial change efforts.

Finally, Part III provides individual participants' perspectives on some of the issues that emerged during the *talk about* sessions. These papers may serve to facilitate discussions regarding issues related to improving the methods and mathematics courses used in the preparation of teachers of middle grades mathematics. Other topics include the way the programs of study and individual courses prepare teachers for the diversity of their learners, address alternative forms of assessment, and model the implementation of technology to enhance mathematical learning. The monograph closes with one participant's reflective examination of the assumptions we make and our need "to keep exploring, listening, thinking, and studying our assumptions."

References

Brown, C., & Borko, H. (1992). Becoming a mathematics teacher. In D. A. Grouws (Ed.), *Handbook of Research on Mathematics Teaching and Learning* (pp. 147-164). New York: Macmillan.

Ferrini-Mundy, J. (1994, February). *Research on the Teaching and Learning of Undergraduate Mathematics*. Seminar presented at East Carolina University.

Flanders, J. (1987, September). "How Much of the Content in Mathematics Textbooks is New?" *Arithmetic Teacher, 35*(1), 18-23.

Fuson, K. (1988). Summary comments: Meaning in middle grade number concepts. In J. Hiebert & M. Behr (Eds.), *Number Concepts and Operations in the Middle Grades* (pp. 260-264). Reston, VA: National Council of Teachers of Mathematics.

Geddes, D., & Fortunato, I. (1993). Geometry: Research and classroom activities. In D. T. Owens (Ed.), *Research Ideas for the Classroom: Middle Grades Mathematics* (pp. 199-224). Reston, VA: National Council of Teachers of Mathematics.

Hiebert, J., & Behr, M. (1988). Introduction: Capturing the major themes. In J. Hiebert & M. Behr (Eds.), *Number Concepts and Operations in the Middle Grades* (pp. 1-18). Reston, VA: National Council of Teachers of Mathematics.

Hilton, P. (1994, February). *Improving the Undergraduate Preparation of Mathematics Teachers: A Mathematician's Perspective*. Seminar presented at East Carolina University.

Kloosterman, P., & Gainey, P. H. (1993). Students' thinking: Middle grades mathematics. In D. T. Owens (Ed.), *Research Ideas for the*

Classroom: Middle Grades Mathematics (pp. 3-21). Reston, VA: National Council of Teachers of Mathematics.

Leitzel, J. R. C. (Ed.). (1991). *A Call for Change: Recommendations for the Mathematical Preparation of Teachers of Mathematics.* Washington, DC: The Mathematical Association of America.

Leitzel, J. R. C. (Ed.). (1994). *Charting Directions for Research in Collegiate Mathematics Education: A Conference Report.* Washington, DC: The Mathematical Association of America.

National Council of Teachers of Mathematics. (1989). *Curriculum and Evaluation Standards for School Mathematics.* Reston, VA: Author.

National Research Council. (1989). *Everybody Counts: A Report to the Nation on the Future of Mathematics Education.* Washington, DC: National Academy Press.

National Science Foundation. (1993). *Proceedings of the National Science Foundation Workshop on The Role of Faculty from the Scientific Disciplines in the Undergraduate Education of Future Science and Mathematics Teachers.* Washington, DC: Author.

Selden, A., & Selden, J. (1993, November). Collegiate Mathematics Education Research: What Would That Be Like. *College Mathematics Journal, 24*(5), 431-45.

Stake, R., Raths, J., St. John, M., Trunbull, D., Jenness, D., Foster, M., Sullivan, S., Denny, T., & Easley, J. (1993). *Teacher Preparation Archives: Case Studies of NSF-Funded Middle School Science and Mathematics Teacher Preparation Projects.* Urbana, IL: University of Illinois, Center for Instructional Research and Curriculum Evaluation.

Steen, L. A. (Ed.). (1992). *Heeding the Call for Change: Suggestions for Curricular Action.* Washington, DC: The Mathematical Association of America.

Steen, L. A. (Ed.). (1990). *On the Shoulders of Giants: New Approaches to Numeracy.* Washington, DC: National Academy Press.

Wagner, S., & Kieran, C. (Eds.). (1989). *Research Issues in the Learning and Teaching of Algebra.* Reston, VA: National Council of Teachers of Mathematics.

Part I

Middle Grades Curriculum Projects

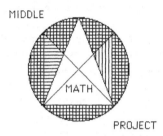

Before the first MIDDLE MATH conference each curriculum project
was asked to write two papers: one paper providing an abstract describing
their project and the other paper addressing the implications of their
curriculum for middle grades teacher preparation programs.

Specifically, they were asked to respond to: What are the characteristics
of a preservice program that would prepare teachers to implement your
curriculum effectively? They were to address the following boldfaced
areas (the questions were provided to facilitate their thinking about the
areas):

Knowledge of Mathematics—What mathematics courses should a
teacher take to be comfortable with your curriculum and to see implications and extensions of it?

Knowledge of School Mathematics—How will pre-service teachers
connect the mathematics they are learning to the mathematics they will
be teaching? What content in your curriculum are today's *prospective*
middle grades mathematics teachers likely *not* to have learned at the
pre-college level and how should they explore this content? In what
ways should they revisit the pre-college content that they have learned?

Knowledge of the Teaching of Mathematics—What kinds of classroom experiences should prospective teachers have to accommodate

the beliefs about instruction reflected in your curriculum? What facility
in using classroom tools such as manipulatives, calculators, and/or
computers is optimal for implementing your curriculum? What modes
of instruction should preservice teachers have experienced in mathematics courses and/or in methods courses?

Knowledge of the Learning of Mathematics—What assessment
techniques should prospective middle grades mathematics teachers
know about and/or be evaluated with to accommodate the beliefs about
evaluation reflected in your curriculum? What should they know about
the research on the difficulties students have in learning middle grades
mathematics? How should they learn about this research?

Knowledge of the Learner of Mathematics—Your curriculum was
designed for use with diverse student populations. How would you
prepare teachers to manage a diverse student population using your
curriculum? What should they learn or experience to facilitate the
inclusion of all kinds of students?

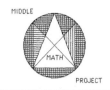
The Connected Mathematics Project (CMP)

Principal Investigators: Glenda Lappan, Elizabeth Difanis Phillips and William Fitzgerald (deceased), Michigan State University; James Fey, University of Maryland; and Susan Friel, University of North Carolina at Chapel Hill

Project address:
A717 Wells Hall
Department of Mathematics
Michigan State University
East Lansing, MI 48824

Phone: (517) 432-2870
FAX: (517) 432-2872

To order materials
please contact:
Prentice Hall Publishers
501 Boylston St., Suite 900
Boston, MA 02116

Intended Grade Levels Served by the Project: 6–8

Abstract:

The *Connected Mathematics Project* (CMP) was funded by the National Science Foundation between 1991 and 1997 to develop a mathematics curriculum for grades 6, 7, and 8. The result was *Connected Mathematics,* a complete mathematics curriculum that helps students and teachers develop understanding of important concepts, skills, procedures, and ways of thinking and reasoning in number, geometry, measurement, algebra, probability, and statistics. In 2000, the National Science Foundation funded a revision of the *Connected Mathematics* materials, CMP 2, to take advantage of what we learned in the six years that CMP has been used in schools. CMP 2, as with CMP, went through a five-year revision process. Each unit went through at least three cycles of reviews, revision, field-testing and evaluation. Forty-nine schools, approximately 150 teachers and 20,000 students were involved in the revisions.

The curriculum, teacher support, and assessment materials that comprise the *Connected Mathematics* program reflect influence from a variety of sources:

- knowledge of theory and research;

- authors' vision, imagination, and personal teaching and learning experiences;

- advice from teachers, mathematicians, teacher educators, curriculum developers and mathematics education researchers; and

- advice from teachers and students who used pilot and field-test versions of the materials.

The fundamental features of the CMP program—focus on big ideas of middle grades mathematics, teaching through student-centered exploration of mathematically rich problems, and continual assessment to inform instruction—reflect the distillation of the advice given and the experience gained from those varied sources.

Connected Mathematics at a Glance

Below are some key features of *Connected Mathematics:*

- *It is organized around important mathematical ideas and processes.* The mathematics in the curriculum is carefully selected and sequenced to develop a coherent, connected curriculum.

- *It is problem-centered.* Important mathematical concepts are embedded in interesting problems to promote deeper engagement and learning for students. Students develop deep understanding of key mathematical ideas, related skills, and ways of reasoning as they explore the problems, individually, in a group, or with the class.

- *It builds and connects mathematical ideas from problem to problem, investigation-to-investigation, unit-to-unit, and grade-to-grade.* The name of the curriculum points to the importance of students making connections among mathematical ideas. Rather than seeing mathematics as a series of unrelated experiences, students learn to recognize how ideas are connected and develop a disposition to look for connections in the mathematics they study—it has coherence.

- *It provides practice with concepts and related skills.* The in-class development problems and the homework problems give students practice distributed over time with important concepts, related skills, and algorithms.

- *It helps students grow in their ability to reason effectively* with information represented in graphic, numeric, symbolic, and verbal forms and to move flexibly among these representations.

- *It supports instruction and learning based on inquiry.* The teacher launches the problem, the students explore the problem individually or in small groups with the teacher guiding, probing, redirecting, extending as needed, and then together the class summarizes the mathematics and reasoning.

- *It is for teachers as well as students.* The *Connected Mathematics* materials were written to support teacher learning of both mathematical content and pedagogical strategies. The teacher's guides include extensive help with mathematics, pedagogy, and assessment. Multidimensional tasks are provided in the assessment materials.

- *It is research-based.* Each *Connected Mathematics* unit was field tested, evaluated, and revised over a six-year period. Approximately 200 teachers and 45,000 students in diverse school settings across the United States participated in the development of the curriculum.

Overarching Goal of CMP

The overarching goal of *Connected Mathematics* is to help students and teachers develop mathematical knowledge, understanding, and skill along with an awareness of and appreciation for the rich connections among mathematical strands and between mathematics and other disciplines. The single mathematical standard that has been a guide for all the CMP curriculum development is:

All students should be able to reason and communicate proficiently in mathematics. They should have knowledge of and skill in the use of the vocabulary, forms of representation, materials, tools, techniques, and intellectual methods of the discipline of mathematics, including the ability to define and solve problems with reason, insight, inventiveness, and technical proficiency.

Student Materials

The mathematical content developed in Connected Mathematics covers number, geometry, measurement, statistics, probability, and algebra appropriate for the middle grades. Connected Mathematics 2 is organized into 24 carefully sequenced units (8 per grade level). One additional unit at each grade level continues to be available from the first edition of Connected Mathematics to meet specific state or local needs. Each unit develops a *big* mathematical idea, that is, an important cluster of related concepts, skills, procedures, and ways of thinking. Each unit provides three to five investigations that contain two to five major problems for students to explore in class. Applications, Connections, and Extensions (ACE) help students practice, apply, connect, and extend their understandings. Investigations culminate in a *Mathematical Reflection*, helping students articulate their understandings and connect *big* mathematical ideas and applications.

Teacher Materials

The CMP materials reflect the understanding that teaching and learning are not distinct—*what to teach* and *how to teach it* are inextricably linked. The circumstances in which students learn affect what is learned. The needs of both students and teachers are considered in the development of the CMP curriculum materials. This curriculum helps teachers and those who work to support teachers examine their expectations for students and analyze the extent to which classroom mathematics tasks and teaching practices align with their goals and expectations.

With any important mathematical concept, there are many related ideas, procedures and skills. At a grade level in CMP a small, select set of important mathematical concepts, ideas, and related procedures are studied in depth rather than skimming through a larger set of ideas in a shallow manner. This means that time is allocated to developing understanding of key ideas in contrast to *covering* a book. The Teacher Guides accompanying CMP materials and related support materials were developed to support teachers in planning for and teaching a problem-centered curriculum. The teacher materials engage teachers in a conversation about what is possible in the classroom around a particular lesson.

- The goals of each lesson are articulated and suggestions are made about how to engage the students in the mathematics task, how to promote student thinking and reasoning during the exploration of the problem, and how to summarize with the students the important mathematics embedded in the problem.

- An overview and elaboration of the mathematics of the unit is included along with examples and a rationale for the models and procedures used. This mathematical essay is to help a teacher stand above the unit and see the mathematics from a perspective that includes the particular unit, but also connects to earlier units, and projects to where the mathematics goes in subsequent units and years.

- Actual classroom scenarios are included to help stimulate teachers' imaginations about what is possible.

- Questions to ask students at all stages of the lesson are included to help teachers support student learning.

- Suggestions are provided for involving students with special needs and those who are English Language Learners.

- Strategies for supporting students' learning through group-work are included.

- Reflection questions are provided at the end of an Investigation to help teachers assess what sense students are making of the *big* ideas and to help students abstract, generalize, and record the mathematical ideas and techniques developed in the Investigation.

- Multiple kinds of assessment and grading help are included to help teachers see assessment and evaluation as a way to inform students of their progress, parents of students' progress, but, also, to guide the decisions a teacher makes about lesson plans and classroom interactions. More demanding partner quizzes are used to mirror classroom practices as well as highlight important concepts, skills, techniques, and problem solving strategies.

- A website for parents provides information on individual units and the lesson problems in a unit. The website supports parents in helping students with homework.

Mathematics in Context: A Connected Curriculum for Grades 5–8 (MiC)

Principal Investigator: Thomas A. Romberg

Project address: Wisconsin Center For Education Research
                        1025 W. Johnson St.
                        Madison, WI 53706

Phone: (608) 263-3605
FAX: (608) 263-3406

To order materials   Holt, Rinehart and Winston
please contact: Order Fulfillment Department
                        6277 Sea Harbor Drive
                        Orlando, Florida 32887-0001

Intended Grade Levels Served by the Project: 5–8

Abstract:

This project has been funded by the National Science Foundation to create a comprehensive mathematics curriculum for the middle grades that reflects the content and pedagogy suggested by the NCTM *Curriculum and Evaluation Standards for School Mathematics* and *Professional Standards for Teaching Mathematics*. The development of the curriculum units reflects a collaboration between research and development teams at the Freudenthal Institute at the University of Utrecht, The Netherlands, research teams at the University of Wisconsin, and a group of middle school teachers.

A total of 40 units have been developed for grades 5 through 8. These units are unique in that they make extensive use of realistic contexts. From the context of tiling a floor, for example, flow a wealth of mathematical applications, such as similarity, ratio and proportion, and scaling. Units emphasize the inter-relationships between mathematical domains, such as number, algebra, geometry, and statistics. As the project title suggests, the purpose of the units is to *connect* mathematical content both across mathematical domains and to the real world. Dutch researchers, responsible for initial drafts of the units, have 20 years of experience in the development of materials situated in the real world. These units were then modified by staff members at the Univer-

sity of Wisconsin to make them appropriate for U.S. students and
teachers.

Because the philosophy underscoring the units is that of teaching
mathematics for understanding, the curriculum will have tangible
benefits for both students and teachers. For students, mathematics
should cease to be seen as a set of disjointed facts and rules. Rather,
students should come to view mathematics as an interesting, powerful
tool that enables them to understand their world. All students should be
able to reason mathematically; thus, activities will have multiple levels
so that the able student can go into more depth while a student having
trouble can still make sense out of the activity. For teachers, the reward
of seeing students excited by mathematical inquiry, a redefined role as
guide and facilitator of inquiry, and collaboration with other teachers
should result in innovative approaches to instruction, increased enthusiasm for teaching, and a positive image with students and society.

The project has developed an entire mathematics program (40 units,
teachers' guides, assessment materials, and staff development materials) with the resources and support necessary for successful implementation. Each of the 40 units will use a theme that is based on a problem
situation that should be of interest to students. These themes are the
"living contexts" from which negotiated meanings can be developed
and sense making can be demonstrated. Over the course of the four-
year curriculum, students will explore in-depth the mathematical
themes of number, common fractions, ratio, decimal fractions, integers,
measurement, synthetic geometry, coordinate and transformation
geometry, statistics, probability, algebra, and patterns and functions.
Although many units may emphasize the principles within a particular
mathematical domain, most will involve ideas from several domains,
emphasizing the interconnectedness of mathematical ideas. These units
are not meant to be used as a text but are designed to be a set of materials that can be used flexibly by teachers, who tailor activities to fit the
individual needs of their classes.

Philosophically, we operate under a social constructivist model. Students will be working individually and in flexible group situations,
which include paired work and cooperative groups. We believe that the
shared reality of doing mathematics in cooperation with others develops a richer set of experiences than students working in isolation.

Students exiting *Mathematics in Context* (MiC) will understand and be
able to solve non-routine problems in nearly any mathematical situation

they might encounter in their daily lives. In addition, they will have gained powerful heuristics, vis-à-vis the interconnectedness of mathematical ideas, that they can apply to most new problems typically requiring multiple modes of representation, abstraction, and communication. This knowledge base will serve as a springboard for students to continue in any endeavor they choose, whether it be further mathematical study in high school and college, technical training in some vocation, or the mere appreciation of mathematical patterns they encounter in their future lives.

When teachers, accustomed to traditional texts/instructional methods, are confronted with true innovations in pedagogy and curriculum, they go through a period of adjustment that can often be frustrating (or even frightening) if they have not been exposed to some level of training and support. In conjunction with the course, the staff development component of the project will involve the creation of multimedia applications that will help teachers work through the initial frustration of practical change and will provide ongoing support as they grow in their teaching practice. Through the use of interactive video/Hypertext applications, teachers can observe model teachers teaching the units they are using currently, can make hypotheses regarding the effectiveness of these methods, and can evaluate their hypotheses with respect to what students learned in the observed lessons. These applications will relate directly to the curriculum units and software; thus, they will provide a direct link to teachers' practices. In addition, more general applications will be developed to address concerns, such as lack of resources, including parents and administrators as partners in the teaching process, and multiculturalism in the mathematics classroom.

Assessment in the MiC curriculum will be treated primarily as an ongoing, informal aspect of the teaching/learning process rather than a special "assessment event." In the in-service program, we will be helping teachers develop observation skills and questioning strategies that help teachers determine what their students are thinking when engaged in a task. In addition, we will be helping teachers gain an understanding of new methods of assessment, such as portfolios, projects, performance tasks, and open-ended questions.

Middle-school Mathematics through Applications Project (MMAP)

Principal Investigators: Shelley Goldman and George Pake, Institute-Research-Learning and Jim Greeno and Ray McDermott, Stanford University

Project address: The Institute for Research on Learning
                       66 Willow Place
                       Menlo Park, CA 94025-3601

Phone: (415) 614-7900
FAX: (415) 614-7957

To order materials: Project—see address above.

Intended Grade Levels Served by the Project: 6–8

Abstract:

When we began the *Middle-school Mathematics through Applications Project* (MMAP), mathematics learning in school was in a crisis: over 90% of the students who began the high school math sequence were not finishing; girls' achievement in math was dropping during the middle school years; inner-city youth were failing in school, and patterns of alienation were firmly reflected in high drop-out rates. Success or failure in school mathematics had the power to qualify or exclude students from future school or employment opportunities. With opportunities for future jobs increasing in the science and technical fields, mathematics became a major gate. Without high school mathematics of algebra and above, students were unprepared for either entrance to college or technical education.

The problem was not that mathematics was too hard for most to learn; we were simply failing our children with our attachment to traditional math instruction. The NCTM Standards had won the battle for reform at the policy level and the battle now needed to be fought at the classroom level. We had to find new ways to help teachers teach and learners learn mathematics. We believed that one way to do that was to reorganize the approach to math pedagogy and content.

Our grant from the NSF allowed us to develop and study the use of materials that experimented with an applied approach to middle school math. We believed that in order to develop a new kind of math curriculum that would make a difference for underserved students it would not be enough to simply develop new classroom materials. We would need to stretch our researcher perspectives to be more inclusive of others' views, challenge standard ways of conceiving how mathematical content gets incorporated into curriculum, and to try to understand and address issues affecting teachers' work in mathematics classrooms. This meant paying attention and working on curriculum, teaching, learning, assessment, and technology.

The challenge of MMAP was to take guidance from research on learning, the concerns of middle school students, and conjectures about approach, content, and activity flow in classrooms in order to create curriculum materials that could work within national constraints for standards and accountability. We made several commitments to—

- Bring to the curriculum design process a collaborative community that included education researchers, teachers and teacher educators, curriculum developers, math-using professionals and students.

- Define and test the feasibility of an applications-based approach to learning math in middle schools.

- Learn more about the ways that technologies might be integrated into the mathematics classroom through a research and development process.

- Create a series of application units and assessments for middle school math classrooms.

- Work with teachers and learn about the issues they face as they make changes in their perspectives and practices.

- Conduct research to improve on our materials design and generate new understandings of mathematics teaching and learning.

We based our curricula on the assumption that no matter what we are doing, we are learning. The question, then, is not "How do we get young people to learn;" rather it is, "What are young people learning in our classes?" In traditional school mathematics, youngsters take on the role of students. In that role, they learn to carry out pre–existing algorithms to solve problems that are carefully matched to the algorithm— while not necessarily understanding the algorithm or the problem.

However, when people use math outside of school, roles are different. This changes the character of the math that is learned, and we believe that middle schoolers can attain greater conceptual understanding of mathematics when they participate in activities like those of people who use math in their work. In the *Guppies* unit, for example, students assume the role of population biologists. Each student plays a role in a group, and a "community" of population biologists is formed. Through group work, shared goals, and discourse, mathematical problems emerge. Because understanding the math can be helpful in solving a real problem, the math becomes *functional*. Discussion about the problems takes place, and new mathematical understanding emerges.

We have built math materials that take into account the social necessities that dominate the classroom and community lives of young adolescents. We have accomplished this for a full range of students, not just the children who achieve in a traditional math program or in the high track classes. In every evaluation to date, our environments have opened opportunities for children who often appear "out of it" to emerge as math learners. Informal documentation of student achievement in MMAP classrooms, where two to three major units are used each year, show positive growth.

MathScape: The Seeing and Thinking Mathematically Project (STM)

Principal Investigator: Glenn Kleiman

Project address: Education Development Center (EDC)
                         55 Chapel Street
                         Newton, MA 02160

Phone: (617) 969-7100 ext. 510
FAX: (617) 965-6325

Email: glennk@edc.org

To order materials   Glencoe/McGraw-Hill
please contact: Order Department—P.O. Box 543
                         Blacklick, OH 43004-0543

Intended Grade Levels Served by the Project: 6–8

Abstract:

The *MathScape*: *Seeing and Thinking Mathematically* (STM) curriculum builds upon the central theme of *mathematics in the human experience*. This was conveyed in the opening paragraph to our proposal:

> To be human is to seek to understand. Mathematics, along with science, has made possible dramatic advances in our understanding of the physical universe. To be human is to explore. Throughout history, mathematics has been essential for exploration, from navigating by the stars to travel into space. To be human is to participate in a society. Societies require mathematics to keep records, allocate resources, and make decisions. To be human is to build, and mathematics is essential for the design and construction of everything from tents to temples to skyscrapers. To be human is to look to the future. Mathematics enables us to analyze what has been, predict what might be, and evaluate our options. To be human is to play, and mathematics is part of our games and our sports. To be human is to think, to create, and to communicate. Mathematics provides a vehicle for thinking, a medium for creating, and a language for communicating. Indeed, to be human is to develop mathematics. Mathematics has been developed in every culture for the purposes of counting, locating, measuring, designing, playing, and explaining.

Through units of the curriculum, students will experience mathematics as it is used for planning, predicting, designing, exploring, explaining, coordinating, comparing, deciding and for other activities fundamental to human endeavors throughout the world and throughout history.

The content of the curriculum includes *processes of mathematical investigation* (e.g., abstracting, representing, generalizing, proving, creating, applying, and communicating); four *central mathematical ideas* (proportional reasoning, multiple representations, patterns/functional relationships, and modeling); and *specific concepts, skills,* and *language* in the areas of algebra, estimation/computation, discrete mathematics, functions, geometry/visual reasoning, measurement, number, probability, and statistics.

The pedagogy of the curriculum reflects a view of learning as a process of constructing one's own knowledge and emphasizes the importance of the social context of learning for middle school students.

Assessment is integrated with learning activities. Our units contain performance assessment opportunities. To do so, we select challenges that seem to provide particularly good windows into what students know and don't know. We provide teachers with suggested scoring rubrics and samples of student work to which we have applied the rubric. Opportunities to use and develop portfolio assessment are provided by projects in which students write, design, and build models.

Technology is integrated throughout the curriculum. In addition to calculators, students use a set of general-purpose software tools we created, e.g., the *Geometry Inventor*. They also use software and video designed to accompany specific units (e.g., a business simulation and video about the design and building of houses around the world).

Collaborating with EDC are school districts in the Boston area, the San Francisco school district, three subcontractors (Shell Centre for Mathematical Education, Education Collaborative for Greater Boston, Inverness Research Associates), a consulting group in Australia directed by Charles Lovitt, and a publisher (Sunburst/Wings for Learning).

MathThematics: Six Through Eight Mathematics (STEM)

Principal Investigators: Rick Billstein, Director
                        Jim Williamson, Chair-Writing Team

Project Address: Department of Mathematical Sciences
                        University of Montana
                        Missoula, MT 59812

Phone: (406) 243-2659
FAX: (406) 243-2474

Web Site Address: http://www.math.umt.edu/~stem/*

To order materials   McDougal Littell
please contact: 222 Berkeley
                        Boston, MA 02116-3764

Intended Grade Levels Served by the Project: 6–8

Abstract:

The *Six Through Eight Mathematics* (STEM) Project is designed to
develop a new mathematics curriculum for grades 6–8, including a
complete set of student materials, teacher support materials, and assessment materials. STEM is designed to provide teachers with curricular materials that are mathematically accurate, utilize technology, and
provide students with bridges to science and other mathematical fields.
The materials are designed to integrate communication into mathematics by providing opportunities for students to use reading, writing, and
speaking as tools for learning mathematics. STEM materials are problem-centered, application based, and use technology where appropriate.
Many lessons are designed to be project oriented and have students
work cooperatively. New assessment techniques are used throughout
the materials. In designing the curriculum, the staff worked cooperatively with IBM, McDougal Littell, Texas Instruments, and Microsoft.
After the national field tests were completed, the materials were published for dissemination during the 1997-98 school year.

Students who complete *MathThematics* will have acquired the mathematical skills necessary to solve problems, to reason inductively and
deductively, and to apply the numerical and spatial concepts necessary

to function according to their needs in a technological society. Students will be independent learners, well prepared for both work and further experiences in mathematics. They will have the knowledge, ability, and confidence to explore mathematics at the secondary level. They will be experienced in working with extended projects, cooperative learning activities, technology, hands-on materials, applications, modeling, and new assessment techniques. STEM students will learn to think mathematically, to become decision makers, and to view mathematics as relevant to their lives and connected to other areas.

Our goal is to convince students that mathematics is exciting and useful. We expect students to learn mathematics by doing mathematics in a variety of settings. Students leaving this program should be very strong in communication skills and solving application problems. Topics, such as quantitative literacy and discrete mathematics, receive much more emphasis than in the present curriculum.

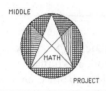
Implications for Teacher Preparation from the *Connected Mathematics Project*

Susan N. Friel
University of North Carolina – Chapel Hill

Knowledge of Mathematics

Preservice teacher candidates who will be teaching middle grades mathematics need to participate in a *mathematics program* that (1) helps them identify and understand the big ideas in middle grades mathematics and (2) provides "standards-inspired" opportunities for learning mathematics that mirror the mathematics learning environments we are trying to create for middle grades students.

a) The mathematics program is highlighted to emphasize the need to view middle grades mathematics teacher preparation as a comprehensive program that promotes coherence and consistency across learning experiences (as opposed to "taking a collection of mathematics courses" to fulfill some set of requirements). Both the institutional perspective and the teacher candidate perspective need to be programs that understand and contribute to a continual development of coherency and direction.

b) The NCTM *Curriculum and Evaluation Standards for School Mathematics* identify a number of content standards for middle grades mathematics and, through discussion, highlight some of the deep ideas that underpin what needs to be addressed in middle grades mathematics. The *CMP* curriculum has a number of major content strands that reflect consideration of the directions set in the *Standards*; within each strand, we have identified the "deep" ideas that we think are important for middle grades students. We would want teachers to be comfortable with these big ideas and capable of working with students to support their mathematics learning.

Knowledge of School Mathematics

We would argue that at least part of the mathematics and pedagogy experiences of teacher candidates needs to be coordinated to support looking at school mathematics from an advanced mathematics perspective. Teacher candidates need to engage at a much deeper level in the subtleties of the mathematics taught in the middle grades. One way to do this is to provide "bridging courses or seminars" for teacher candidates that are tied to their college mathematics courses—so they can talk about what mathematics is central to the development of middle grades students' mathematical knowledge, including the mathematics taught in elementary and secondary school programs. Teacher candidates need a broad enough perspective so that they are able to extend their thinking from consideration of one mathematics unit to a consideration of a year's program of mathematics, and finally, to a consideration of what constitutes the mathematical life of a student K-12.

Knowledge of the Teaching of Mathematics

Teacher candidates need well-developed "habits of mind" to think about how to teach mathematics. They need opportunities to examine big ideas in mathematics and ways of teaching the big ideas. Then they need to explore what it means to develop problems that will raise the issues that we want students to confront in order to promote deeper thinking about mathematics in the middle grades. They also need experience with all sorts of tools, coupled with an in-depth examination of the ways these tools help make mathematics accessible to students. There is enough research now that shows students don't readily move back and forth between many of the more formal manipulative models and the mathematics being represented by these manipulative models. Rather, we want students to use more general tools, e.g., cubes, polystrips, geoboards, calculators, chips, and so on, that let them represent and explore their own questions in their own ways.

Teacher candidates progress through their own stages of concern and growth during their preservice programs. Many of us recognize that teacher candidates are better able to think more deeply about mathematics learning after their student teaching experiences. Ideally we would want a sequence of learning experiences for teacher candidates that takes advantage of these stages. One model might be—

- A pre-student teaching "methods" course that addresses their needs for survival (e.g., How do I set up and run a mathematics class?) and support.

- Student teaching with ongoing support and mentoring that is responsive to the day-to-day experiences of the teacher candidates.

- A final mathematics course with an associated pedagogy seminar that analyzes what is happening in the mathematics course in terms of instruction and learning and assists teacher candidates in developing tools that help make them better able to reflect on their own teaching.

An ideal arrangement would be to maintain a connected relationship with teacher candidates after graduation as they complete their first three years of teaching (the induction period), working with them in an extended program that continues to involve them in learning mathematics and reflecting on their own practice of the teaching of mathematics. This would continue consideration of "what are the big ideas in mathematics with which they need to be concerned" and address the need for helping teacher candidates build added depth of understanding over time.

Knowledge of the Learning of Mathematics and of the Learner of Mathematics

We have not done a very good job in helping teacher candidates learn how students learn mathematics. We need to address the theoretical/psychological perspectives of how students (particularly adolescents) learn and ways teachers teach. Middle grades teacher candidates need a rich understanding of the nature of middle school students. Clearly, this responsibility extends beyond the mathematics and/or mathematics education departments.

In addition, we need to orient teacher candidates' learning toward an action research perspective, one in which they are involved in a number of different ways — and for a number of different purposes — in finding out how students think and learn mathematics. For example, teacher candidates have studied a given mathematical idea and have looked further at its relevance within school mathematics. Using problems they have designed, they follow-up these experiences with interviews or other work with middle grades students with the intention of better understanding what the students know. This establishes a different role for a teacher—one as a person who listens carefully and asks questions in order to understand what the learner understands.

Another strategy linked to action research employs the use of cases as a tool for promoting discussion of teaching and learning. There are materials available that focus on mathematics at both the middle (e.g., Barnett, 1994) and secondary (e.g., Silverman, Welty, & Lyon, 1994) levels. In addition, a casebook by Shulman and Mesa-Bains (1993) is now available for teachers and teacher educators that specifically considers questions of diversity in the classroom. The cases in these books are not merely narratives or descriptions of an event or series of events; rather, they are "cases of something." The reader must actively engage in the context provided. Both reading cases and taking on the challenge of writing cases address the need to help teacher candidates become more focused on teaching and learning in specific contexts.

References

Barnett, C. (Ed.) (1994). *Fractions, decimals, ratios, and percents: Hard to teach and hard to learn?* Portsmouth, NH: Heinemann.

Silverman, R., Welty, W. M., & Lyon, S. (1994). *Teaching methods cases for teacher problem solving.* New York: McGraw-Hill.

Shulman, J. H. and Mesa-Bains, A. (1993). *Diversity in the classroom.* Hillsdale, NJ: Lawrence Erlbaum Associates.

Implications for Teacher Preparation from *Mathematics in Context*

Meg R. Meyer
Wisconsin Center for Education Research

Knowledge of Mathematics

Preservice teachers would benefit from a strong precalculus curriculum that would include, but not be limited to, trigonometry, discrete mathematics, matrix algebra, geometry (emphasis on connections rather than proof), elementary functions, probability, and quantitative literacy. The focus of these courses should be on—

- Concept understanding rather than symbol manipulation.

- The connectedness of mathematics rather than on isolated concepts and skills.

- Multiple representations and solution strategies.

- The experience of mathematics as growing out of and being applied to everyday situations.

Whatever curriculum the preservice teacher experiences, the attitudes and habits they take away from it are of (at least) equal importance to the mathematical content. Their experience of mathematics should lead them to believe that—

- Mathematics is a sense-making activity at which they can be successful; because and since their success is not exceptional, they in turn can expect their students to be successful.

- Mathematics is connected—they will look for the connections and not be satisfied with the learning of isolated ideas and disembodied skills.

- There are a variety of different approaches to a given problem and often there are different answers, depending on initial assumptions and interpretations.

- Real mathematics problems take longer than two minutes to solve; false starts are the norm and not a reflection on their ability— persistence pays off.

- They can always make a start on a problem, even if they don't know where it will lead or if it will be successful.

- Learning and doing mathematics are intertwined activities that continue throughout our lives.

Knowledge of School Mathematics

It is unlikely that preservice teachers will have experienced geometry from the MiC approach. This approach can be characterized as "look and see" geometry with a focus on what you see and how you see it. The best way to explore this content is to do the mathematics of the geometry units as though they are students. This means really trying to answer the questions without looking in the teacher's guide and discussing answers and strategies with other teachers. These attitudes and methods are important even when the content is familiar, in order for teachers to see the development of ideas and connections to other topics in mathematics.

Another important difference between MiC and their own experience of middle school mathematics is the place of mastery. It is usually not expected that students will master concepts and skills with the first encounter. Instead, mastery is achieved over time and usually over the course of several units and grade levels. Without this understanding it will be difficult for teachers to "trust" the curriculum, and they will be tempted to supplement it with more explanation problems and the dreaded worksheet.

Knowledge of the Teaching of Mathematics

Whenever possible, preservice teachers should learn mathematics and methodology in classrooms that model the pedagogy we would like to have used with MiC. Such a classroom would—

- Be built upon notions of social constructivism.

- Show flexible use of groups. Cooperative groups is only one of a number of different groupings that are possible. Teachers need to understand the decision making process behind the various choices.

- Model successful management of materials, so that manipulatives, calculators, etc. are as common place and available as paper and pencils.

- Model classroom discourse in which communication is encouraged and multiple strategies are valued.

- Model what it means to be a learner and a teacher at the same time.

- Provide preservice teachers with the opportunity to see kids solve problems without instruction, so that they can begin to learn that kids can do much more than we think.

- Challenge the beliefs and attitudes of preservice teachers and engender critical self-reflection.

Knowledge of the Learning of Mathematics

Preservice teachers should be given the opportunity to understand assessment primarily as a way to gather information to inform instruction, and secondarily as a way to assign grades. This shift in emphasis encourages and values informal methods of assessment like observations and interviews. Any field component should require preservice teachers to interview students in order to learn how to ask questions, listen to and interpret responses, and plan instruction as a result. They should have experience with journal writing and portfolio assessment. They should have assignments that are purposely open and ill defined so that they can experience the need to state assumptions and make choices. They should be given feedback and the opportunity to make revisions to the work-in-process.

Knowledge of the Learner of Mathematics

Here again the attitudes and beliefs of the preservice teacher are critical. They need to believe that mathematics for all is an obtainable goal. We continually underestimate what kids can do, if given the chance. This attitude is clearly related to teachers' own experiences as learners of mathematics. Most teachers probably learned mathematics in an environment that more closely resembled a "filter" than a "pump," and they, as a result, likely see their own role as a teacher as being a part of that filtering system. This belief needs to be exposed and challenged repeatedly. Teachers need to think about adapting their curriculum and pedagogy to meet the needs of diverse students. A field component that exposes preservice teachers to students different from themselves seems to be a minimal response to meet this need.

Implications for Teacher Preparation from the *Middle-school Mathematics through Applications Project*

Rick Berg
Institute for Research on Learning

MMAP units are based on a new way of thinking about mathematical applications. Traditionally, applications were something that came at the end of the chapter (in the form of a few of those dreaded word problems) where students practiced the algorithm they had just learned to do. Rather than a list of mathematical skills to be learned, each MMAP application is a "slice" of the real world that has potential for engaging young people in mathematics. Students work within a scenario with a problem or provocative question that encourages them to create and analyze designs. These scenarios provide ample opportunities for mathematics to be used and learned along the way.

Our working hypothesis has been that students encountering math opportunities within a project they care about are more likely to want to learn mathematics than students wading through a string of math lessons. Additionally, they will have evidence for their own mathematical arguments in their designs and analyses. Research in math classrooms so far supports these hypotheses.

"Design" is the vehicle that MMAP uses to get students engaged in activity. As students work through cycles of research, analysis and design, they will encounter many places to use and learn mathematics. Some of these cycles are built into the curriculum, others are unpredictable; for example, in the *Codes, Inc.* unit, a student may begin to focus on the differences in the graphs of quadratic and cubic functions. We provide handouts to help students through some of the predictable math opportunities, but the unexpected opportunities are a chance for the teacher to question students, set up new activities, conduct a class discussion, or provide a reference. This places responsibility on the teacher to recognize and act upon these opportunities for math learning.

In general, preservice education teaches geometric and algebraic concepts at a level that may not be particularly helpful for middle school students. Teachers should spend some time thinking about the things they take for granted when engaged in mathematical activity:

What do I have to know to use a scale on a map? What different aspects
of functions are highlighted by graphs or tables? Answers to these
questions are generally not investigated at the preservice level, although they are important concepts for middle-school math. One way
of getting teachers to think about these issues is to have them use our
materials. We have found from many inservices that teachers can learn
a lot about math, as well as about teaching, from participating in one of
our curriculum workshops.

The kinds of activities we suggest as "characteristics of a preservice
program that would prepare teachers to implement our curriculum" are
not yet commonly available to—or easily organized for—preservice
and inservice teachers alike. Classes, such as industrial arts (or other
design electives), history of math, logic, and perhaps even statistics,
probability, and number theory, would be considered electives to be
taken beyond the requirements of the typical credential program. These
courses, along with algebra and geometry, would be helpful for teachers preparing to use MMAP materials. Some might view this program
as an added burden on teaching candidates. One institutional shift that
would effectively support preservice teachers would be to change the
orientation of required courses or to offer alternatives to the typically
required sequence of courses.

One alternative is to offer experiences outside of the classroom. In
MMAP we have sent teachers into the workplace to look for math.
Teachers worked with professionals in many settings, including product design firms, architectural offices, scientific laboratories, and fire
stations. They develop new perspectives on how to teach math to
prepare their students for the future and on what math is. Thinking
about "what math is" is fundamental from our perspective. Math looks
a lot different in context than it does in traditional textbooks or classrooms. Students bring many different mathematical perspectives to the
classroom, not always the "standard" ones. Part of our commitment to
diversity rests upon the teacher's ability to recognize the different
forms of contribution that kids bring to the table.

Another aspect of teacher preparation that is often neglected is helping
the teacher learn to prepare. We've found that each time a teacher uses
project-based curriculum, such as a MMAP unit, some new facet of
preparation emerges. Teachers *discover* what they need to know to
teach the unit effectively. The design of teacher preparation programs
should cater to this need. How do teachers find the kinds of resources
they need to support their work? Resources can include any of the

following: people who can act as mentors for students and the teacher, practicum sites for teachers, events that support the needs of teachers (school-world and "real" world), materials, time, etc. Considering our list of characteristics, this kind of "just in time" response to teachers' preparation needs might be worth promoting—both as an orientation to teacher preparation and as an activity people in education can learn to support.

In addition to content, teachers should have experience using various processes in the classroom. All of our units are designed to be collaborative. Teachers should know strategies for structuring and maintaining group work. Teachers should understand the design process, be able to formulate a problem, and cycle through research, design, and analysis processes to arrive at a solution. They should be able to connect math content to the real-world issues which provide the context for our units. As previously noted, they should be able to recognize different versions of math in the classroom and formalize them if necessary. Teachers should recognize that complex problems can be solved in multiple ways, with multiple mathematical approaches.

Preservice teachers should also be given access to technology. We have found that teachers generally are more comfortable using MMAP materials if they know the basics of the Macintosh computer: opening documents, saving, cutting and pasting, moving windows, etc. Teachers should have enough of a background in technology to recognize the appropriate tool for the task the students are doing. Technology should be a tool for learning, not an obstacle.

There are certain attitudes educators should have in a MMAP classroom. MMAP is not a lock-step, content-driven curricula; teachers should be able to take risks. The teacher's authority is based in leadership, not in having the answers to all of the problems. Many times students are in a better position to answer questions than the teacher. Teachers need to be able to ask authentic questions in design contexts in order to make sense of the sensible things students are doing. Teachers should also learn the role of facilitator. When students get stuck during an activity, pointing out a resource in the classroom (like a meter stick) is often a better step for the teacher than a lecture might be.

Assessment should be ongoing in MMAP units. The teacher is always trying to keep up with the students, rather than the other way around. Teachers should understand that assessment should reflect the activity in which the students are engaged, and it should reflect the content of the learning as well as the context of the learning. It should be used, not only to assign grades, but to see improvement and plan next steps. Teachers should be comfortable listening to their students—peer assessment and self-assessment are important in MMAP classrooms. Students should be given many opportunities to show the teacher what they know. Students should also be able to show their understanding using a wide range of media (text, graphs, tables, pictures, oral presentations, posters, etc.).

Teachers should be aware of current research in education. In many cases, however, the research focuses on students who are *not* learning, and why that might be the case. In MMAP units, it is sometimes just as hard to see a student *learning*. Research is important to help teachers see learning in complex environments such as MMAP units.

Implications for Teacher Preparation from *MathScape*

Susan E. Janssen
Education Development Center

An Example

I would like to start with an anecdote about a significant experience in my own background as an undergraduate math major headed for a career in teaching. The anecdote illustrates one of the key elements necessary for the preparation of mathematics teachers who will be implementing new curricula. Headed for a career in teaching, I took a Number Theory course from David Burton at the University of New Hampshire. Beginning the second day of class, we were asked to write homework solutions on the board for the class to discuss. During the first week, few volunteers were forthcoming so Dr. Burton drafted "volunteers." Reluctantly, we would slink up to the board, write out our answer and sit back down hoping there was little to criticize in our work. To my surprise and delight, Dr. Burton did not focus on criticism, but instead asked us to explain what we had done and then asked the rest of the class what they thought about the response. Invariably, a discussion of mathematics would ensue. Our responses were viewed as work in progress, and over the first few weeks, Dr. Burton built an atmosphere in the classroom in which it was acceptable and safe to work with incomplete ideas, to be confused (and even wrong) because that was part of learning to think about number theory. I finally had something to contribute to the creation of the mathematics we were doing. We became a small community of mathematicians, doing number theory, and it was rich, exciting and fun. By the end of the course, my classmates and I were no longer slinking up to the board. Instead, we were enthusiastically volunteering to write up problems with which we had struggled because class had become a place to think about and explore mathematics.

Learning to Do Mathematics

Helping prospective mathematics teachers develop a sense of what it means to do mathematics in a community of mathematicians may be the most important element necessary for their preparation as they undertake new curricula, such as *MathScape*. Our curriculum addresses functions, graphs, creating and manipulating algebraic expressions, 2-D and 3-D geometry, probability and statistics, topics in discrete math, number systems, and making mathematical arguments. Ideally a preservice teacher would have majored in mathematics and have studied all these areas. However, a teacher who understands what it means to explore and think about mathematics in one area can apply this kind of thinking to other areas of mathematics. Thus, it is more important to help teachers develop a sense of doing mathematics than it is to make sure they complete course work on a defined set of topics.

Changing the Culture of College Classrooms

I chose the example of Dr. Burton's Number Theory class because it reflects the kind of classroom culture needed in college courses as well as middle school classes. New curricula ask a lot from teachers—to engage students in open-ended investigations that relate mathematics to a variety of human experience; to help students think and learn about "big ideas" in mathematics, such as proportional reasoning or multiple representations; to conduct discussions; to draw out student understanding; to assess learning in new ways. In defining their own teaching style, educators draw heavily on their own experience as a learner. Thus, preservice teachers need to experience what it is like to learn in this way, to know what their students will be experiencing. College courses need to present teachers with problems and investigations which invite them to explore the mathematics, not simply to find the expected response. Professors need to draw out their students' thinking, and therefore, will need to understand more about how learning and understanding is constructed in order to present their students with problems and explorations which build deeper understandings.

The implications of the classroom culture described above are significant to the college and university community. The current "reform movement" is part of an ongoing evolution that must also extend to college classrooms that are moving away from a lecture format and toward an active, engaging format in which students build and construct their own understandings.

To emphasize and extend the changing educational culture, preservice teachers can be assessed in a variety of ways—through interviews, writing, and presentations, as well as quizzes and tests. They would understand the criteria by which they will be evaluated before they are actually evaluated. Once they understand the scope of the task on which they will be judged, they should be involved in setting up the assessment criteria.

Teaching in a Changed Culture

Many preservice teachers will face the challenge of teaching students with a wide variety of racial backgrounds, using curricula that cannot possibly represent all cultures equally. One solution is to develop materials that are "culturally open," in which any student can bring his or her own experience to bear on a task—no matter what his or her racial or ethnic background is. For example, students can use geometric concepts to design and build a house that fits a particular climate. Or students can learn about number systems and number composition by looking for mathematical patterns in the words used for numbers in different languages. Preservice teachers should engage in similar kinds of activities in which they can put something of themselves into the mathematics they are learning. In turn, they should search for and try to create activities for middle school students in which the mathematics is tied to a human context and is relevant to the lives of students from any background.

Diversity appears in many forms in the classroom, including diversity in students' ability levels. Teaching a class that contains many different ability levels has never been easy, and it gets more challenging as heterogeneously-grouped classrooms become more popular. To address this issue, one of the basic tenets of the *MathScape* materials is that mathematical investigations are designed to be accessible to all students. Activities and investigations are written so that all students can find an entry point, get started, and then can take problems to differing levels of complexity (according to each student's ability). Preservice teachers must also have experience with investigations that provide an entry point for anyone, and that can be taken to various levels of complexity. For example, in one *MathScape* unit, students use a circular device with beads to create their own number system and explore its parameters. In another unit, students conduct probability experiments with coins to create a class collection of "strip graphs"—rectangular strips of paper with squares colored red or yellow to indicate heads or tails—as a way of examining probability intuitively. In both examples, any student can begin work on the investigations at a surface level, and go on to explore them with greater mathematical sophistication.

Becoming a Reflective Teacher/Practitioner

At an NCTM Conference in the early 1990s, I had the opportunity to hear Judith Sowder of San Diego State University speak about her work with "excellent" teachers, trying to identify what they had in common. The one element they shared was a deep reflectiveness about their own practice. If teachers are to use new curricula well, they need to learn ways in which they can be reflective about their practice. To support this need, *MathScape* includes several features that encourage teachers to reflect on the mathematics and on their students' learning of the mathematics. The *MathScape* curriculum is designed to be something that grows with a teacher. This goal is accomplished by providing a default sequence for the first-time user while writing the materials in a way that encourages the teacher to make decisions about the entire curriculum. Mathematical background pieces in the materials describe layers of connections and understanding that can be brought out as the teacher becomes more familiar with the materials through repeated use. *Teacher Reflections* are brief narratives from field test teachers, describing teachable moments that occurred in their classrooms and their reflections on each. A teacher who has learned ways to be reflective can make the most use of this aspect of the materials.

Preservice teachers can learn tools for reflecting on their own practice in their college preparation. They can be encouraged to read current research in mathematics education as part of their class work, and to think about if and how they would utilize the research results in their own classroom. Written reflections on case studies can form the basis for class discussions on different aspects of teaching and learning mathematics. Preservice teachers can interview students working on a mathematics problem, and then analyze the students' thinking.

Implications for Teacher Preparation
from *MathThematics*

Rick Billstein
University of Montana

Within the last fifteen years, the National Science Foundation (NSF) has funded twelve mathematics curriculum projects. Three of the elementary (K-6) projects are commercially available. Five middle school projects followed the elementary projects and are also available. Four secondary projects were then funded and were available a few years after the middle school projects. With this effort, NSF has laid the groundwork for substantial K-12 curriculum reform in the 21st century.

These new curricula are significantly different from traditional programs. The new programs focus on improving mathematics education for *all* students. Mathematical knowledge for all students is important because future students will be living in a more technological and scientific world. Although this paper will focus on the changes at the middle school level, most of the comments are true for all levels.

Future middle school teachers will be asked to teach different mathematics in different ways using different technologies and materials. The future success of this reform movement depends on the inservice training of teachers in the field as well as changing the teacher preparation of future teachers. The last major reform movement, often referred to as the "New Math," had a major effect on publishers. However, teachers and future teachers received little training in the new programs and eventually moved "back to the basics" because they felt more comfortable with the basics. The "New Math" was much more successful at the secondary level because the majority of secondary mathematics teachers received some sort of training (such as a summer institute). At the K-8 level, this training did not happen and the reform was not well received. This is a lesson that must be learned in order for the new curricula to have lasting impact. The following descriptions pertain to the Six Through Eight Mathematics Project (STEM) and address how the training of teachers will need to change for this project to succeed.

Knowledge of Mathematics and School Mathematics

STEM materials are problem-centered, application based, and use technology where appropriate. Many teachers will face mathematics content that they have not seen before. The content strands for the project are—

- Mathematical Reasoning, Problem Solving, and Communication
- Number
- Measurement
- Geometry
- Statistics
- Probability
- Algebra
- Discrete Mathematics.

In addition, four unifying concepts have been identified that are relevant to almost every strand:

- Proportional Reasoning
- Multiple Representations
- Patterns and Generalizations
- Modeling.

The mathematical strands provide the breadth for the curriculum and the unifying concepts provide the depth. Future teachers need work in both areas—the unifying concepts must also receive heavy emphasis. Most teacher preparation programs prepare teachers for the number related topics, but geometry, statistics, probability, algebra, and discrete mathematics are often neglected.

Not only is much of the STEM content new, but it is introduced in an integrated, application-based format. STEM is designed to be taught in thematic modules that last approximately 4 weeks each. Themes include topics such as *Search and Rescue, Mathematics at the Mall,* and *Comparisons and Predictions.* STEM is interdisciplinary in nature with strong ties to science, social science, and literature. Teachers need to make connections within and outside of mathematics. Teaching for mastery the first time a topic is introduced is no longer the norm and teachers must learn to trust that the topic will be adequately covered throughout the year. In an integrated program, topics are covered numerous times in different settings and at different levels.

To become familiar with the content in the STEM program, future teachers should take a minimum of one full year of a *Mathematics for*

Teachers sequence that covers all the topics previously mentioned. The courses should be taught by experienced mathematics educators — too often these courses are taught by teaching assistants, part-time faculty, or mathematicians with no interest in mathematics education. The year-long sequence should be followed by or taken in conjunction with a methods course and the content and methods courses could be integrated. If possible, this year long content sequence should be followed by an additional course or two that are chosen depending on the content of the year-long sequence. Typical courses such as *Mathematics for Liberal Arts* or *College Algebra* are not acceptable for preservice teacher training. Students in these types of courses are usually exposed to a technology-void curriculum that has little to do with teaching middle school mathematics. The teaching that usually takes place in these courses does not exemplify what is needed for future teachers.

Knowledge of the Teaching of Mathematics

In STEM, the role of the teacher changes from that of a lecturer to that of a facilitator, listener, questioner, and prober. The lecture is an acceptable way of teaching in some cases, but it is only one method that future teachers need to see in their program. Through the questions they ask and the classroom atmosphere they create, teachers become discussion leaders who provoke students' reasoning about mathematics. Teachers in many cases become co-investigators with their students in learning and doing mathematics. Students operate on many different levels of thinking — teachers must be able to recognize and judge these levels and use them to teach their classes better. Future teachers should be exposed to this kind of atmosphere in the classes they are taking.

The STEM curriculum is more than a textbook. It is made up of activities, hands-on materials, software, and hardware, such as scientific fraction calculators and graphing calculators. Teacher preparation should include experiences with manipulatives (including pattern blocks, multi-base blocks, geoboards, MIRAs, compasses, protractors, geometry models, and algebra tiles). Future teachers should use scientific fraction calculators and should be familiar with graphing calculators. They should have some experience with computer software tools, such as a spreadsheet, a geometry utility, a statistics package, and some current educational software. At some point they should also be exposed to the Internet and the World Wide Web.

Knowledge of the Learning of Mathematics and of the Learner of Mathematics

Students quickly learn that what is assessed is what is valued. No matter what progress is made in changing mathematical content or in teaching techniques, if assessment practices are not changed, then a new curricula will not work the way it was designed. Embedded assessment, authentic assessment, performance assessment, and self–assessment are all important in the STEM project.

Open questions and scoring rubrics play a major role in the STEM project. A multidimensional general scoring rubric is introduced in the sixth grade and is used for the three years. Future teachers need training with new assessment techniques, including work with scoring rubrics. Books, such as NCTM's *Assessment Standards for School Mathematics,* should be read and discussed by the class. Self-assessment should be a part of the courses that future teachers take and journals should be kept to record progress. Teacher behavior should be modeled and discussed in class. Discussions should include the number and type of questions asked in class; the amount of time spent on various activities, such as teacher presentation, class discussion, group work, and seat work; the length of student responses; the handling of cooperative groups; and the scoring of student work.

The STEM materials were commercially available through McDougal Littell/Houghton Mifflin during the school year 1997-98. Other middle school reform projects appeared about the same time. It is important that the training of future middle school teachers be addressed now and that these future teachers be ready for the reform curricula that will be available when they graduate. As one sixth grade pilot teacher explained it, "This is a whole new approach to teaching. We are like co-investigators and we have to present this program in a way that the kids can solve problems, instead of presenting answers."

In summary, to prepare future teachers for the new curricula and teaching styles, preservice courses must be redesigned and taught by appropriate role models. The new curricula are becoming available, but teachers with proper backgrounds are needed to make them work in the way that they were intended.

Part II

Case Studies

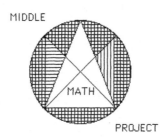
MIDDLE

MATH

PROJECT

Each MIDDLE MATH University was expected to revise or develop, in the course of the first summer conference and the 1995/96 academic year, a model for the middle grades teacher preparation program at their institution. They also had to modify or develop and test at least one course to be used in the preparation of middle grades mathematics teachers. At the second conference, the representatives from each participating institution were to present a session on the changes they had made in their middle grades program. They had three options for their presentation.

The first option was a poster session. Groups of 5-6 listeners would rotate through the posters every fifteen minutes so that the presenters could speak to each group about what they had done. The second option was a formal presentation of about 30 minutes. Presenters could speak on the course(s) they had modified, the structure of their middle grades program, or a specific topic in teacher preparation. The final option was for the presenters to facilitate a "talk about" session on a topic of particular concern.

Rather than summarize the participants' sessions for the monograph, participants had opportunity to write a case study that highlighted some of the ways they were attempting to improve the preparation of teachers of middle grades mathematics. Participants submitted proposals and a variety of cases were selected based on the size and type of program involved. Here are some examples of what MIDDLE MATH participants accomplished. If you are looking for a specific topic within the case studies, the following table may be helpful.

Case Study Matrix

Topics \ Institutions	East Carolina University	Elon College	Florida International University	Lees McRae College	Mercer University	The Ohio State University	Saint Xavier University	University of North Carolina at Chapel Hill	University of South Florida	University of Wisconsin-Oshkosh	Western Carolina University	Western Michigan University	Wright State University
Whole program	■		■			■	■			■	■		■
Detailed description of course	■				■	■							
Assessment									■				
Field experience			■			■		■				■	
Integrated curriculum						■							
Mathematical reasoning	■												
Multicultural issues			■	■									
Pedagogical model	■			■		■				■	■	■	
PRAXIS exam				■									
Small college		■		■									
Student beliefs and attitudes		■				■		■					
Technology	■	■	■										■
Use middle grades curricula	■		■					■	■		■		
Algebra course	■									■			
Geometry course		■											
Intro. course with applications		■											■
Methods course							■		■				■
Modeling course										■			
Pre-Calculus course											■		
Problem solving course					■								
Probability & statistics course	■									■			■

EAST CAROLINA UNIVERSITY

**Rose Sinicrope, Mary Eron, Sid Rachlin, Ron Preston,
Cheryl Johnson, Mike Hoekstra, and Sunday Ajose**
East Carolina University

The Institution

East Carolina University (ECU) is a comprehensive public institution
that is part of the University of North Carolina system. Located in
Greenville in eastern North Carolina, ECU enrolls nearly 18,000
students, making it the third largest university in North Carolina and
the largest teacher training institution in the state. In fact, East Carolina
boasts the 15th largest education program in the United States, and with
minority students totaling nearly 14% of total enrollment, ECU is the
11th largest preparer of minority educators nationally.

With ten faculty members specializing in mathematics education, ECU
includes one of the largest mathematics education units in the country.
The mathematics education faculty are part of the thirty-five member
Department of Mathematics, which also includes statistics, computer
science, and pure mathematics. Mathematics education faculty have
opportunities to teach mathematics at the graduate and undergraduate
levels, as well as content and pedagogy courses specifically designed
for the preparation or continued development of teachers.

The Program

As of 1992, North Carolina was one of eight states to issue a middle
grades certification or license endorsement. In 1994, ECU decided to
revise its middle grades mathematics program to make it more in tune
with the needs of middle grades teachers. The MIDDLE MATH Project
was created to further that effort. By 1996, a new program was designed and put into place in which students selecting a specialty in the
teaching of mathematics at the middle grades level complete a 27
semester hour minor called the Math Concentration. The concentration
includes a three-credit-hour mathematics methods course which requires ten-clock-hours of field experience.

The Math Concentration can be sub-divided into three levels. The
initial level consists of *Algebraic Concepts and Relationships*, *Data*

Analysis and Probability, and *Euclidean Geometry*. All middle grades majors must take the first of these courses and at least one of the other two. *Algebraic Concepts and Relationships* and *Data Analysis and Probability* are new courses whose development started with the MIDDLE MATH Project. The *Euclidean Geometry* course had already been revised to be more interactive and to reflect the recommendations for reform proposed in the NCTM *Standards* (1989, 1991) and the MAA *A Call for Change* (Leitzel 1991).

The second level of ECU's Math Concentration includes those courses which form the body of the program for middle grades majors selecting mathematics as one of their two concentrations. These courses include *Pre-Calculus Concepts and Relationships*, a two-semester sequence in *Calculus for the Life Sciences*, and *Discrete Mathematics*. In revising our program, we investigated new approaches to the teaching of mathematics at the middle grades level and at the college level, as-sessed our course offerings, and identified mathematical needs of local middle grades teachers. We found it necessary to combine aspects of two existing courses: a trigonometry course and a pre-calculus course. The result was a new course: *Pre-Calculus Concepts and Relationships* that, with a heavy reliance on a graphing calculator, explores logarith-mic, exponential, and trigonometric functions and uses matrix algebra to solve problems. The calculus sequence provides our concentration students with their only opportunity within the program to interact mathematically with non-education majors. It also provides an under-standing of the many real-world applications of mathematics. The discrete math course is an introduction to logic, sets, mathematical induction, and matrix operations. Applications within the course are drawn from probability, linear programming, dynamical systems, social choice, and graph theory.

In addition to the methods course, the third level of the concentration includes the capstone course, *Applied Mathematics via Modeling*. This is a new course that will provide a "look back" at the mathematics in the middle grades program through the vehicle of mathematical model-ing. Projects done in a true modeling (that is, problem-solving) spirit will be used throughout the course. Thus, students can expect to use any mathematical content, technique, or strategy from their background to formulate a solution to a project. The intent of *Applied Mathematics via Modeling* is not to introduce new content, but to find applications for previously introduced concepts that will deepen preservice teachers' understanding of that content.

The Accomplishments

Algebra

In this part of the case study, we focus on the course *Algebraic Concepts and Relationships*, including its development, and the pilot of the course in the spring of 1996. *Algebraic Concepts and Relationships* was designed to replace College Algebra as the introductory level course taken by all prospective middle grades teachers. The purpose of the course is to have students take a reflective look at algebraic systems to deepen understanding of algebraic properties and interconnections. Rather than use a textbook, students work on problem sets and new topics are introduced through class discussion of problems. Thus, the course focuses on the processes students use to learn to solve problems. The assignments are designed to promote the development of conceptual understanding over time. Each topic is developed over the course of several problem sets. Each problem set includes problems from middle grades student materials, real world applications, skills development, analysis of student errors, a reading assignment that is often tied to research on student learning or research on teaching, and a writing or journal assignment (see Appendix at the end of this chapter). Students were encouraged to keep all their work in a notebook that they submitted at the end of the semester along with a portfolio demonstrating their progress in the course.

A team of four instructors developed and taught the first offering of this course: Dr. Mary Eron, a mathematician; Ms. Cheryl Johnson, a middle grades teacher; Dr. Sunday Ajose and Dr. Sid Rachlin, both math educators. Collaboration between members of the developmental team was essential in designing drafts of problem sets that guided the instruction. Eron, Ajose, and Rachlin taught the classes in daily or weekly succession. This is important, not as a model for curriculum development, but because it forced the team to discuss how to present problems in class, what mathematics could be brought out using the problems and, after class, to reflect on what had been accomplished and how it could have been done more effectively. We felt these interactions were invaluable in designing the course. Each class was video taped to help the instructors reflect on the design of the instructional model and the course materials. Students were interviewed at the end of the semester to aid us in understanding their reactions to the course.

There is general agreement that prospective teachers should have a thorough knowledge of the mathematics they are going to teach, as well as some idea of where that mathematics is going or how it can be used as a foundation for later development. The topics of number systems and algebra fit into that way of thinking. In *Algebraic Concepts and Relationships*, we arranged the topics in a somewhat unique fashion. We organized the course around algebraic operations, rather than the individual number systems, and then looked at operations within the number systems. The operations we considered were equivalence, addition/subtraction, multiplication/division, and power/root—choosing not to look at whole numbers, integers, rational numbers and real numbers separately. We also included complex numbers and polynomials. Students saw the same operation in a variety of situations and made connections between operations using real numbers and operations in algebra. The syllabus also included a topic on functions. We stressed interpreting functions, graphing functions and then interpreting the graphs and solving equations.

In designing the problem sets, we chose to emphasize the NCTM *Curriculum Standards* of problem solving, reasoning, connections, and communication and the following four *"big ideas"*—(1) reflection on acquired knowledge, (2) ties to the middle grades classroom, (3) tools for learning, and (4) research on mathematics teaching and learning.

Reflection on Acquired Knowledge

Problem Solving: We began the course with problem solving to emphasize that this is the process through which all mathematics is done. Students were given a problem set to do at home. The next class was spent discussing the problems, interweaving old and new topics from knowledge that students already possessed. Occasionally, we gave an impromptu lecture on a topic most students did not understand. The students were then asked to present their work, both orally and in writing, giving us a chance to observe their mathematical reasoning.

We assumed students had seen the content of the course before and that they would be taking a reflective look at the material, emphasizing the ties between the various concepts. We believed that if students were thinking about their knowledge of algebra in relation to other mathematical concepts that they "knew" and were possibly reconstructing this knowledge, they would be more likely to recall the needed ideas in later courses or teaching situations.

Reasoning: The students experienced a lot of frustration at being asked to do problems on material we had not discussed or problems that had not previously been modeled for them. For example, one student stated in her evaluation of the course, "This course would be much more useful to me if the problems were explained before I was asked to do them for homework. I never felt like I was grasping any of the concepts covered in class." Prompted by this feedback, we now make part of the class a forum on how frustration is a natural part of doing mathematics. Frustration is something to be coped with rather than done away with—in the real world, one would not be given problems for which the solutions were already known.

We did provide hints in the class to deal with frustrating problems. We looked at specific examples or smaller cases and, in general, we tried to make sure that students understood what the problem was asking. With the *locker problem*, for example, we had the students start over three times before they were ready to present their solutions. One of our best students wrote in her portfolio that "when a problem is as much a challenge as this was, nothing can compare to the feeling of accomplishment one gets by finding the solution and being able to explain the solution and how it was achieved." She also included her attempts at the problem in the portfolio. The first attempt was scratched out. The second time, she got closer to the solution. She tried a third time and succeeded. Then she was able to get up in front of the class and convince everyone that hers was the solution to the problem and why her solution worked. It became evident that it is not always the teacher's role to provide solutions. Sometimes a sense of direction is all a student needs to solve a problem on his/her own.

Connections: Peter Hilton, a prominent mathematician, advised us as we began the MIDDLE MATH project, "It is not so important what mathematics students learn, but the attitude toward mathematics that they acquire." If students have a positive attitude, they will be willing to learn more mathematics in school or on their own. This idea is especially important for teachers. Statements made in student interviews attest to the fact that the course was successful in creating positive attitudes. One of our weaker students was a music education major taking this course to fill his math requirement. In his portfolio, he wrote about a conversation he had. Someone said to him,

We, as choral directors, are problem solvers. We hear a problem with a sound and try to correct it with our knowledge and skills. That is what math and music are about. They are one and the same...Only through gaining knowledge and sharpening skills will I become a better choral director. The same is true for

math. Only through gaining knowledge and the sharpening of
my skills will I become a greater critical thinker. I will leave
[the course] not only as a better mathematician, but a better
musician because another door has been opened for me.

Communication: Students were asked in their writing assignment to
reflect on what they learned each week. These journals turned out to be
a useful tool. For example, some students wrote about how important,
as future teachers, mathematics vocabulary was. They found that in this
course not only did they have to solve mathematical problems, but they
had to talk about them—so the need for good mathematical vocabulary
became apparent. Early on in the course, one student had this to say:
"When I'm placed in the teacher's position, I become more aware of
my vocabulary. I begin using the appropriate vocabulary when explaining the problem, and I think that it's important to know how to use math
words and how to use them to convey a message to students."

Another student wrote something similar in her portfolio. She felt this
course combined math and English because, as part of their homework
assignments, they had to write journals, do projects, and create portfolios. Again, she said, "Almost any math problem of length, especially
word problems, will contain vocabulary—vocabulary like monomial,
binomial, exponent, rational numbers, and radicals, to name a few.
With a firm understanding of these terms, almost any word problem
can be—if not solved—at least understood." She moved into a reflective mode where she looked at problems, not just as something to solve,
but as something to explain. We did not talk about the use of vocabulary in the class, yet students realized that in order to use mathematics
effectively, they had to be able to communicate ideas successfully so
that other people could understand what they were trying to say.

Ties to the Middle Grades Classroom

We felt strongly there should be direct connection between the mathematics preservice teachers are learning and the mathematics they would
be teaching. Thus, we tried to tie our course to the middle grades
classroom in several ways. First, we included problems from student
materials on the problem sets. These problems were often taken from
the curriculum projects funded by the NSF described earlier in the
monograph. Some of the problems were used for concept development.
Others were used as reinforcement, and they were referenced so students knew where they came from. We think that it was beneficial for
students to do the problems and then talk about why and how they

might be used rather than having students simply read about the problems—as is often done in texts designed for education majors.

Second, we assigned a project in which students had to compare and contrast student texts. They were given a topic, a traditional text and an innovative text from one of the NSF curriculum projects, and asked to write about the similarities and differences between the two texts. They were also asked to compare the texts to the North Carolina Standard Course of Study. The level of sophistication in the students' analyses varied greatly. Although they saw some advantages to the innovative texts, they were much more comfortable with the step-by-step procedures found in the traditional texts. The project gave the students a chance to use some of a classroom teacher's standard references as well as observe how curriculum is changing and how expectations of teachers will change with it.

The team's middle grades teacher offered her insights on what middle grades students would think or do. She also taught a highly interactive *workshop* session which included activities like an algebra walk (where the students were points on a line on the sidewalk) and a relay game (where the students turned written expressions into algebraic symbols).

Tools for Learning

"Prospective mathematics teachers learn about pedagogical content knowledge when their instructors model activities; introduce tools, such as manipulatives and technology; and discuss literature about how students learn certain mathematical concepts and about student misconceptions" (MSEB, 1996, pp. 6-7). We bought into the idea that teachers teach how they themselves were taught and that students need to see a model that differs from a lecture where the teacher simply imparts knowledge to the learner. We moved toward a model where students learn from each other with a teacher acting as a facilitator. Organizing the students into small groups helped make them actively involved in the class. We took each student from where they were and tried to draw them into solving the problem at hand. Often, we asked students to present their work in front of the class, and because they worked with and got to know other people in the class, they were not intimidated — they saw everyone made mistakes. In fact, in the interviews, students said they liked to go to the board and wished we had done more board work as the material became more difficult.

We also tried to make learning a more hands-on experience than in the traditional college classroom. We made use of graphing calculators and spreadsheet programs to talk about functions and equations. We worked with the pebble model for integers, algebra tiles and lab gear, base five blocks, and fraction bars. We drew pictures for students to use on the problem sets at home and had the actual manipulatives in class to work within groups. Students explored the advantages of using concrete materials. By using these materials, we were also able to draw even the weakest students into the center of class discussion. One student stated

> *With the usage of positive and negative marbles and positive and negative rod units, I now have a more physical interpretation of solving equations. Giving a tangible reference to such an obscure concept has helped me focus on the purpose of solving equations. By giving me a clearer understanding of this old concept, I feel better prepared to teach this concept. This new concept of teaching will surely take the dreariness out of the mathematics classroom.*

Research on Student Learning and Research on Teaching

"Educational research, about K–12 mathematics teaching and learning and about the preparation and continuing education of teachers, is a critical component in the improvement of mathematics teacher preparation" (MSEB, 1996, p. 2). Often the readings for a problem set came from publications that translate research for use in the classroom. For example, articles from *Research Ideas for the Classroom: Middle Grades Mathematics* (Owens, 1993) and *Review of Educational Research* (Parker & Leinhardt, 1995) were used. Occasionally, we stopped in a lesson and reflected on the pedagogy that was just used in the discussion or presentation of material. Teachers use different techniques to get their point across. Often, students do not realize these are techniques that they themselves can learn. To have directly experienced an example of a teaching technique in a content course is more valuable than simply reading or watching the same example.

On several occasions, we had students make up their own problems. They wrote word problems to illustrate different models of subtraction and division. We had them solve a problem and then create a similar problem themselves to work on their ability to use the reversibility of mathematical operations. Another type of problem included in the problem sets was one in which a middle grades student's solution for a math problem was provided, and our students were asked whether there was an error and, if so, how they would correct the student's work. Our students liked playing teacher—discovering and correcting their fictitious middle grades students' mistakes. Often these fictitious problems were part of the writing assignment, so the students had to express their observations and corrections in writing.

In summary, we found that (as always) when you add something new to a course, there is never enough time to accomplish all your goals. However, using a reflective approach, we felt that we did not have to teach every topic from scratch. Also, we assumed that problems assigned (but not gone over in class) became part of a student's knowledge base, and we occasionally tested them on this material.

The students found that it was not enough to be able to solve a problem—they needed to communicate and explain their solution. This realization on the students' part, together with the positive attitude that the students developed toward mathematics, provides evidence that it was worth the time needed to make the course more interactive.

Data Analysis and Probability

In this part of the ECU case study, we focus on a second course that we started to develop as part of the MIDDLE MATH Project. *Data Analysis and Probability* is a sophomore level mathematics course for middle grades education majors. The course goals are for students to develop conceptual understanding of elementary probability and statistics and to encounter experiences that will promote the use of appropriate and sound instruction for middle grades students. In addition, the course is expected to prepare students for their senior-level, action research project.

Three ECU faculty worked as a team on the initial development of the course—Rose Sinicrope and Ron Preston are mathematics educators and Mike Hoekstra is a statistician. The course was first taught during the spring semester of 1996. All three teachers attended class meetings, contributed to the planning of instruction, reflected on teaching and student learning, designed materials and assignments, and evaluated student products. Hoekstra provided most of the instruction. We are committed to continuing with the development of the course and hope to obtain support in the future for another jointly focused opportunity to work on development.

An elementary statistics text (Freedman, Pisani, Purves, and Adhikari, 1992) was critical in the development of the course. The text is noteworthy, among statisticians, because of its distinctive conceptual approach. An underlying principle of the text is the premise that understanding one concept thoroughly is better than understanding many concepts superficially. The use of mathematical formalism is virtually non-existent; for example, the arithmetic mean is always the "average." Prose descriptions are used rather than algebraic formulas; for example, the formula used for computing the correlation coefficient is "r = average of (x in standard units) $\times$ (y in standard units)" (Freedman et al., p.124). Concepts are introduced through examples that use statistics to answer important questions in health care, education, economics, politics, and other fields. The exercises in the text are integral to students' development of the concepts. Exercises are typically of two types. Either the data sets in the exercises are so small that computation can virtually be completed mentally, or they are descriptive questions relating to the presented examples.

We explored experimental design, descriptive statistics, correlation and regression, probability, chance variability, sampling, and confidence intervals. Highlights of the text include histograms based on class intervals of differing widths and a vertical density scale; an extensive treatment of descriptive correlation and regression, including the use of the regression method to predict percentile ranks; and a single model of the random selection of numbered tickets from a box as the foundation for inference topics.

Topics relevant to middle grades teachers not covered within the text are stem plots, box plots, and simulations as a tool for answering interesting, but theoretically advanced, probability questions. We addressed these topics through two student projects that linked the instruction to the middle grades classroom.

For the first project, we asked students to design a controlled experiment and an observational study that could be conducted in an eighth grade classroom. Our students were to describe how they would discuss treatment and control groups, randomization, confounding factors, the extent that the experiment is *blind,* and association and causation with their hypothetical class. After students had an opportunity to read the written directions and think about possible contexts, we had an in-class discussion in which students shared their ideas. The project was a difficult task for our students—in a course consisting of eight students and three instructors, most students met with at least one of us outside of class to discuss their plans and a few students met with all three of us. While advising students, we found that we were all focusing on different aspects of the project, thereby making the project more difficult for some students.

Our study of the projects revealed that students had very surprising misconceptions of association versus causation and of what constituted confounding factors. In their written responses to questions concerning association versus causation, the students did not remain focused on their experiment or study but instead would write about relationships connected to their topics. This was also somewhat typical of their responses to test and final examination questions, despite their success-ful, in-class analyses of studies that claim causal relationships. For several of the students, the difficulties of quantifying variables or obtaining accurate measurements were *confounding factors*. On the first test, we asked students to identify a possible confounding factor in a hypothetical study relating children's reading skills and the amount of television they watched. A simple answer would have been gender. However, not a single student gave an answer that completely captured the idea that *confounding* relates to an inseparable mixture of possible causes of an effect. We asked for a definition for *confounding* on the final examination and still received vague responses.

In our initial planning, we decided that the first project would focus on statistics and the second on probability. The vehicle for the second project was the *Chocolate Chip Cookie Experiment* (Channell 1989). After some concern about the extent to which activities in the article lead students to specific strategies, we decided to give students copies of the labs with some general written directions, which included vari-ous ways to generate a random whole number between one and twelve. Students worked on the project for several weeks with intermittent due dates to share collected data and to discuss their explorations of stem plots and box plots. Sample lessons from the *Connected Mathematics Project* and *Mathematics in Context* (middle grades curriculum projects presented at the first MIDDLE MATH conference) were used to explore these kinds of data displays.

About half the students in our pilot class had children in middle grades. Thus, in addition to their field experiences, our students had access to middle grades texts and had opportunities to interact with middle grades mathematics teachers. Their children were using texts that did not include stem plots and box plots yet, according to the state curriculum, interpretation of stem plots is a fourth grade competency, and the construction of box plots is a seventh grade competency.

Midway through the project, we gave an additional assignment requiring theoretical solutions to related questions. Two of the questions were:

1.   Suppose a batch consists of two cookies. If 8 chips are dropped at random into the two cookies, what is the chance that both cookies have at least 1, 2, 3, or 4 chips?

2.   Drop chips at random until both cookies have at least 1 chip. What is the chance that it takes a total of 2, 3, or 4 chips?

It was easy to supplement instruction with current events—frequently sharing news clippings and cartoons. Statistics packages were used to generate graphs and other supplements to instruction. We had to control our tendency to *add* to the class in order not to overwhelm students— especially considering the limited instructional time available. We often went through several revisions before agreeing upon an activity. For example, one of our top students had difficulty with a textbook exercise dealing with chance. The exercise described a large group of contestants, each with a well-shuffled deck of cards. Each contestant would be dealt the top two cards from the contestant's deck. The student had difficulty with the question of what fraction of the group would be dealt the king of hearts as the second card. Each of us was certain that we could clarify the student's misconceptions. We all tried a different approach, but the student continued to interpret the exercise as a conditional probability, and at one point, appeared to have convinced a number of students his interpretation was correct. We concluded that the example used had too much "noise." This experience heightened our concern about selecting optimal examples for exploring probabilistic concepts.

We believe we were successful in meeting several of our objectives. Students developed an appreciation for the discipline of statistics as distinct from mathematics and *number crunching*. Students left the class with an appreciation of the law of averages and the central role of the normal distribution. They were proficient in the construction of stem plots, box plots, histograms, and scatterplots; they were skillful in interpreting tables and graphs; and they were successful in evaluating statistical reports for methods of data collection and sound inferences.

Issues and Challenges

Although we based the design of *Algebraic Concepts and Relationships* on having students reflect on what they already knew, we found that time for reflection needed to be structured into the course in order for that reflection to be sustained. This structured time for reflections will provide an opportunity to discuss the day's readings, which often dealt with how to teach a particular concept. While we did try to stop in the midst of a lesson to share the decisions we were making in deciding how to teach a topic, these occurrences were random and on the spur of the moment. We are still investigating how to make time for reflection a regular and planned occurrence in the course.

In reviewing our pilot of *Data Analysis and Probability,* we found—

- We need to be more selective in our development of probability.

- We need to find a way to include tests of significance.

- We are still questioning the role of data collection—first for its value and then for an efficient way to manage the collection process.

- We are still seeking a balance between *clean* examples for concept development and real world examples.

In reviewing these pilot courses, we also found we needed to integrate the use of technology into the courses from their very beginning. On the one hand, we feel a nagging concern about our students' limited use of technology. On the other hand, we feel students need to see technology as a problem-solving tool rather than an answer key.

In addition to altering instruction, we want to assess students' understanding of concepts. We feel fortunate to have video tapes of all our classes and a convenient facility to video tape student interviews. In addition to using the curriculum research and development model described by Rachlin earlier in this monograph, we want to offer the course to practicing middle grades teachers, to validate the content from their perspective. Our efforts are clearly a *work in progress.*

References

Channell, D. E. (1989). Problem solving with simulation. *Mathematics Teacher*, 713-716.

Freedman, D., Pisani, R., Purves, R., & Adhikari, A. (1992). *Statistics* (2nd ed.). New York: W. W. Norton.

Leitzel, J. R. C. (Ed.). (1991). *A call for change: Recommendations for the mathematical preparation of teachers of mathematics*. Washington, DC: Mathematical Association of America.

Mathematical Sciences Education Board (MSEB). (1996). *The preparation of teachers of mathematics: Considerations and challenges* (A Letter Report). National Research Council: Center for Science, Mathematics, and Engineering Education.

National Council of Teachers of Mathematics. (1989). *Curriculum and evaluation standards for school mathematics*. Reston, VA: Author.

National Council of Teachers of Mathematics. (1991). *Professional standards for teaching mathematics*. Reston, VA: Author.

Owens, D. T. (Ed.). (1993). *Research ideas for the classroom: Middle grades mathematics*. Reston, VA: National Council of Teachers of Mathematics.

Parker, M., & Leinhardt, G. (Winter 1995). Percent: A Privileged Proportion. *Review of Educational Research, 65*, 421-81.

Rachlin, S., Matsumoto, A., & Wada, L. A. (1992). *Algebra I: A process approach*. Honolulu: Curriculum Research and Development Group, University of Hawaii.

Appendix: Sample Problem Set

Algebraic Concepts and Relationships Problem Set #6

Reading   Schifter, D., & Fosnot, C. T. (1993). A Sample Mathematics
Lesson for Teachers: Xmania. In *Reconstructing Mathematics Education: Stories of Teachers Meeting the Challenge of Reform*. (pp. 41-62). New York: Teachers College Press.

Please attempt all problems

1. Reprinted with permission from Rachlin, S., Matsumoto, A., &
Wada, L. A. (1992). *Algebra I: A Process Approach*. (p. 61 #3).
Honolulu, Hawaii: Curriculum Research and Development Group,
University of Hawaii.

   Two ways to use the pebble model to represent the number –4 are

   and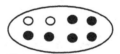

   The first diagram could be described by equations. For example,
–4 + 0 = –4 or 0 + (–4) = –4. The second diagram could be described as 2 + (–6) = –4 or –6 + 2 = –4.

   a. Show two other ways to represent the number –4.

   b. Write two equations for your representations in 1a.

   c. Show two ways to represent the number 3.

   d. Write two equations for your representations in 1c.

2. Find 3 numbers that belong in each region of the diagram below.

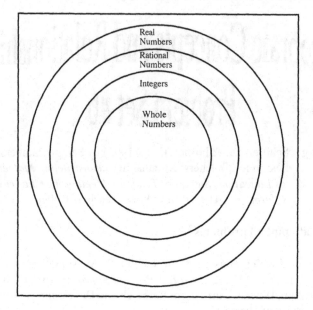

3. Stan spilled ice cream on his homework. He could no longer read the equivalent expressions he had written. Help him fill in the missing parts.

a. $\dfrac{21}{\bigcirc} = \dfrac{6}{10}$

b. $\dfrac{3}{\bigcirc} = \dfrac{33}{\bigcirc}$

c. $\dfrac{\bigcirc}{7} = \dfrac{4}{\bigcirc}$

d. $\dfrac{2}{7} = 28\bigcirc$

e. $\bigcirc = .77\overline{7}$

f. $\bigcirc = .6\overline{23}$

g. $.5 = \bigcirc \%$

h. $.5\% = \bigcirc$

i. $\dfrac{9}{\bigcirc} = 45\%$

j. $\dfrac{2}{3} = 6\bigcirc \%$

4. Adapted from Post, T. R. (Ed.). (1988). *Teaching Mathematics in Grades K–8*. (p. 60). Boston: Allyn and Bacon.

   John is constructing a recreation room in his basement. He has put up the walls and put down a floor. He needs to buy baseboard to put along the walls. The room is 21 feet by 28 feet.

   a. The baseboards come in 10-foot and 16-foot lengths. How many of each kind should he buy?

   b. If John wants to have as few seams as possible, how many of each size baseboards should he buy?

   c. If John wants to have as little waste as possible, how many of each size should be buy?

   d. If the 16-foot boards cost $1.25 per foot and the 10-foot baseboards cost $1.10 per foot, how many of each kind should he buy if he wants to spend the least amount of money?

   e. There is a sale on the 16-foot baseboards. They now cost $0.85 per foot while the 10-foot baseboards still cost $1.10 per foot. How many of each should he buy if he wants to spend the least amount of money?

5. Adapted from Tsuruda, G. (1994). *Putting It Together: Middle School Math in Transition*. (p. 41). Portsmouth, NH: Heinemann.

   The seventh grade class at a certain middle school decided to have an ice cream party to celebrate their outstanding grades. They decided to make ice cream cones, but they were all very independent thinkers, and they each wanted to have a different cone.

   Unfortunately, the school was only willing to provide them with two ice cream flavors, vanilla and chocolate. They knew that if they used only single-scoop cones, they could make only two different cones:

They figured out that if they used double-scoop cones, they could make four different cones:

There are 256 students in this outstanding 7th grade class. How many scoops of ice cream would have to be used in each cone in order to make a different cone for each person?

6. Reprinted with permission from Rachlin, S., Matsumoto, A., & Wada, L. A. (1992). *Algebra I: A Process Approach*. (p. 57 #1). Honolulu, Hawaii: Curriculum Research and Development Group, University of Hawaii.

Bob made a number line out of string with tabs attached to represent numbers. He put his number line in a paper bag and took it to school. When he got to school, he found that some of the tabs had fallen off. Draw these tabs for all of the following numbers to put them back on the number line without straightening it out.

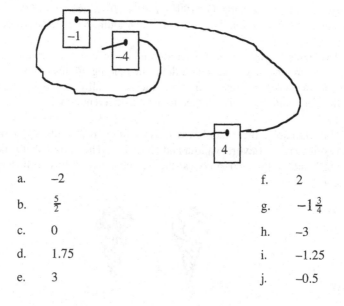

a.	-2	f.	2
b.	$\frac{5}{2}$	g.	$-1\frac{3}{4}$
c.	0	h.	-3
d.	1.75	i.	-1.25
e.	3	j.	-0.5

7. Adapted from *Addison-Wesley Secondary Math, An Integrated Approach*, p. 111) Here are three diagrams and the expressions they represent.

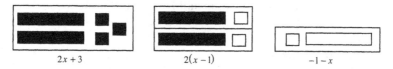

| $2x + 3$ | $2(x - 1)$ | $-1 - x$ |

Draw diagrams to match each expression:

a. $x + 1$ b. $1 - x$ c. $3x - 2$ d. $3(x - 2)$

8. In the reading for tonight it was suggested that you use objects to help you follow the discussion about the creation of a number system for the land of Xmania.

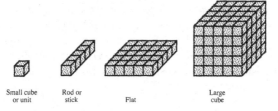

Small cube or unit   Rod or stick   Flat   Large cube

Part I   How would you represent the following amounts using the smallest possible number of objects?

    a. Twenty-three

    b. Fifty

    c. Six hundred twenty-four

    d. Six hundred twenty-five

Part II   Represent each of the amounts in Part I another way (i.e., not using the smallest possible number of objects.)

Writing Assignment #6: To write a fractional representation for a rational number as an equivalent decimal, you can just divide the denominator into the numerator. What do you do to write a given decimal representation for a rational number as a fraction?

Journal Reflection: What have you learned this week that's new and worth remembering? What do you wish you understood better?

Elon College

Janice Richardson and F. Gerald Dillashaw
Elon College

The Institution

Elon College is a small, private, four-year, church-related, liberal arts institution focusing primarily on quality undergraduate education. Teacher education has been a prominent part of the college from its beginnings in 1887. Today, teacher education is the third largest program at the college.

A faculty member with a degree in mathematics education and appointment in the mathematics department coordinates the mathematics education program and serves on the Teacher Education Committee. She is responsible for teaching the middle grades and secondary mathematics methods course and for supervising middle and secondary mathematics student teachers. She also teaches mathematics content courses.

The Program

The program for the preparation of middle grades teachers of mathematics reflects the vision of the *NCTM Standards* and is in line with the North Carolina Department of Public Instruction's Standard Course of Study and teacher competencies. The courses that make up the mathematics concentration for the middle grades program are *Functions with Applications, The Nature of Mathematics*, *Calculus I* and *II*, *Mathematics for Elementary and Middle Grades Teachers*, *Elementary Statistics*, and *Materials and Methods of Teaching Middle Grades Mathematics*. The unique features of the program include small class size and varied field experiences. One of the advantages offered by a small college is the individual attention students receive. In both content and methods classes, preservice teachers receive personalized attention that focuses on their future careers. Students also have the advantage of being in several different middle grades classroom situations. Prior to student teaching, students have field experiences in three courses: *Introduction to Education with Practicum, Curriculum and Instruction in the Middle Grades*, and *Methods and Materials for Middle Grades and Secondary Mathematics*.

The Accomplishments

Elon College has recently made several changes in its program for preparing teachers of middle grades mathematics. These changes have been (1) the deletion of a two-hour math reasoning course and the addition of a four-hour course, *The Nature of Mathematics*, (2) increased use of technology in the teacher education and mathematics programs, and (3) more emphasis on geometry in *Mathematics for Elementary and Middle Grades Teachers.*

A new four-hour liberal arts course, *The Nature of Mathematics*, was added to the middle grades mathematics concentration in the fall of 1996. It replaced a course on mathematical proof techniques. Several ideas came together in the decision to make the change. The first consideration was how best to meet the state competencies for middle grades teachers. Second, ideas shared at the MIDDLE MATH conference were influential in making the change. Many of the new curricula for the middle grades presented at the conference were application-based, and many of the teacher preparation programs presented there stressed mathematical modeling. It was felt that the new course would ultimately prove more relevant to future middle grades teachers.

The new course is rich in up-to-date applications in the social sciences, management sciences, population growth, and statistics. There is heavy emphasis on technology, especially in the area of statistics where students use their TI-82 graphing calculators to deal with raw data in graphing histograms, box plots, and line graphs and in calculating standard deviations, quartiles, means, and medians. In the social sciences, students study and reason through four types of voting methods: plurality method, Border Count method, plurality-with-elimination method, and method of pair-wise comparisons. They discover that different election methods result in different winners. Other applications in the social sciences include weighted voting systems and several fair division schemes. In the management sciences, students explore routing problems, applying their knowledge of Eüler circuits to find the most efficient way possible to route the delivery of goods or services to an assortment of destinations. In the area of population growth, students study linear and exponential growth problems using both recursive and explicit descriptions. At present, we are using the second edition of Peter Tannenbaum and Robert Arnold's text, *Excursions in Modern Mathematics.*

The publisher (Prentice Hall) will provide, at the request of the instructor, student copies of "Themes of the Times" (a supplement of the *New York Times*) containing recent articles which pertain in one way or another to the content of the course. By using these supplements, students are able to see current applications of mathematics in the real world. We have students write summaries and reflections of these articles and then discuss them in class. This greatly increases the students' communication skills in mathematics. At the same time, they realize the potential usefulness of what they are studying.

In response to recent changes made by the state of North Carolina, which include the implementation of a computer skills curriculum for grades 3-8 with an exam given at the end of eighth grade, the teacher education program at Elon College has significantly increased the amount of technology instruction for future teachers. All students are required to take *Computers in Teaching* — a course designed to prepare students to meet essential computer skills required by the State Board of Education. Topics covered in this course include basic computer operations, word processing, databases, spreadsheets, presentation software (i.e., PowerPoint), multimedia software (i.e., HyperStudio), computer-assisted instruction software, and other basic computer applications.

In addition to this *general skills* course, each of the specialty areas has focused on specific technological applications for teaching. Within the Middle Grades Mathematics program, we greatly increased our use of technology as a teaching tool. All courses in the mathematics concentration require the use of a calculator. For the calculus classes, the Hewlett Packard 48G is required; TI-82 graphing calculators are used in the *Functions*, *Statistics*, and *Nature of Mathematics* classes. Additionally, students in the methods classes are using *Geometer's Sketchpad* in lesson planning. Even though manipulatives, such as geoboards and pattern blocks, are excellent aids in visualizing concepts in geometry, students find *Geometer's Sketchpad* especially useful for making conjectures and investigations in transformational geometry.

Because the students do not take a course in geometry, the geometry components in the existing courses needed to be highlighted. In addition to the use of *Geometer's Sketchpad* in the methods course, more emphasis has been placed on geometry in *Mathematics for Elementary and Middle Grades Teachers*. Half the semester is spent covering topics in geometry and measurement in this course that constitutes a survey of the content strands taught in elementary and middle school mathematics. Materials from the Portland State University's *Geometry Notebook* are used in the course.

Issues and Challenges

Through our participation in the MIDDLE MATH Project, we have been able to increase our understanding of middle grades mathematics issues and concerns. Our work in this project has enabled us to revise the middle grades mathematics program to strengthen mathematics content applicable to middle grades. Gaining knowledge of several middle grades mathematics projects has given us multiple perspectives on the approaches to teaching middle grades mathematics and preparing middle grades mathematics teachers.

There was much discussion at the conferences over the appropriate place for technology in a middle grades teacher preparation program. We are fortunate to be able to take a "both/and" approach. We have both a separate technology course *and* applications of technology integrated throughout our content and methods courses. We find that our students benefit greatly from this approach.

There are advantages and disadvantages of being a small school. One problem with which we continue to struggle is that middle grades preservice teachers have no identity of their own. They take their content courses with elementary students and methods course with secondary students; there are not enough students to offer a course for middle grades majors.

Additionally, it is difficult to find good master teachers with which to place student teachers. Knowing about the new middle grades curricula enabled us to find and utilize teachers who are teaching with one of these new curricula. We have been able to place some students with teachers using the STEM materials—a successful and productive partnership between mentor and student teacher has been created that would not have been possible without the MIDDLE MATH Project.

Florida International University

Edwin McClintock and Zhonghong Jiang
Florida International University

The Institution

Florida International University (FIU) is an urban, multi-campus, doctoral-granting, public university in the State University System of Florida. With about 32,000 students, it is a major center of international education and prides itself on the cultural and ethnic diversity of its students and faculty. Located in metropolitan Miami, it serves the fourth largest school system in the U.S.

Faculty with degrees in mathematics education are housed in the College of Education. They are responsible for all aspects of the teacher preparation programs (including teaching methods courses, supervising student teachers, and integrating the use of technology into program courses). They are also responsible for the design of mathematics content courses but seldom actually teach these courses.

The Program

As of fall 1996, no middle school mathematics education program existed at FIU. Rather, middle school math was part of our secondary teacher preparation program. The secondary mathematics education program is strong and growing, but it is substantially different from an appropriate middle school program needed by both the university and the urban school district of southeast Florida. A large number of prospective teachers have an interest in middle school teaching (with many of these teachers desiring to teach mathematics or computer education). Creating a separate middle school program for these individuals has increasingly become a task of top priority. As part of our commitment to the MIDDLE MATH Project activities, we began to develop this program.

The philosophy of the middle grades mathematics degree program is that preservice teachers who complete the program should have substantial breadth across mathematical subfields — in preference to greater depth in fewer mathematical subjects. Therefore, they are asked to pursue 24 semester hours of mathematical sciences. These courses tend to be the first courses in sequences, such as *Calculus I*, *Statistics I*, *Discrete Mathematics I*, *Computer Programming I (Pascal)*, *College Geometry*, *History of Mathematics*, etc. In these courses the use of technology is emphasized wherever appropriate.

The separate middle school mathematics teacher preparation program we are developing has two key features. First, it will be substantially technology-based, concurrently providing middle school certification in technology education as well as mathematics. The program's second feature is exposure to practical experiences. Each year (and possibly each semester of each year) preservice teachers will be involved in teaching middle school students mathematics with technology. Some of the field experiences will take place in an on-campus middle school program, as well as through ties to the public school system.

The Accomplishments

In fall 1994, a program called the Partnership in Academic Communities (PAC), a collaboration between Dade County Public Schools and FIU, was established on the university campus. This program included an "on campus" magnet-like school with students from the district attending daily classes in mathematics, science, and technology. It started as a middle school program involving one seventh grade class and has gradually developed so that currently there are three classes — grades 7–9. Eventually, this program will grow to include individual classes for each of the grades 7 through 12, with one new class being added each year. It is a model that is intended to provide a mechanism for experimentation in meeting the scientific needs of underrepresented, at-risk, minority students (for whom most urban school systems are being only modestly successful as needs providers).

This program's structure also provides our preservice mathematics teachers with frequent opportunities for field experience and practice in implementing the use of technology. The on-campus school uses *Standards*-based curricula and *Standards*-based instruction with at-risk students it serves. Thus, undergraduates can observe university professors and master teachers from the Dade County Public Schools teaching at-risk students using reform-minded methods.

Since the Fall of 1995, our juniors and seniors have been required to participate regularly in the PAC program—working in the computer laboratory and the mathematics classroom each semester. Through the program, our undergraduates experience a view of education that is both demanding and rewarding. *Mathematics in Context: A Connected Curriculum* (MiC) is used in mathematics classes daily. Our prospective teachers' experiences with this curriculum let them experiment with and study (in special focus sessions) the way a *constructivist* curriculum operates and the reactions of middle grades teachers and students to the curriculum.

Related technology classes taught to middle school students make use of *Geometer's Sketchpad* (GSP) and *Microsoft Excel*. The activities in these classes are designed not only to complement the experiences in MiC but to experiment with visual reasoning and investigative forms of learning as well. For example, a major unit the middle school students experience deals with geometric transformations, specifically studying tessellations. As a culminating project for his/her unit on geometric transformations, each middle-school student creates a dynamic Escher construction using GSP (see for example, the figure on the right). The dynamic aspect of GSP extends the work of Escher so the real-life sketches change size, move in an animated fashion, or are proportionally varied. Through observing and assisting in these kinds of activities, our prospective teachers see how middle school  students construct their mathematical knowledge. At the same time, teachers are able to see and investigate how youngsters learn with technology and get a first-hand knowledge of that technology's effect on their attitudes, values, and learning experiences.

In addition to "participating" with a master teacher and university professors, preservice teachers also assume responsibility for instruction one day per week, creating lessons, working with groups, and using technology to develop concepts and problem solving/reasoning skills. This gives them opportunities to see how to integrate technology into the teaching of mathematics and to apply the technology they are learning. This field experience component will allow prospective mathematics teachers to develop a math and technology certification at the middle school level while seeing and experiencing the applications of their learning in a very real and practical way.

Other than the requirement of participating in the PAC program activities, we require each undergraduate preservice teacher to visit public school middle grades classrooms at least five times in the junior year, and at least ten times in the senior year (before student teaching). They are asked to write reports and attend discussions about these visits. One of the tasks assigned to them is to observe whether a class involves the use of technology in teaching and learning mathematics and, if so, how that technology is used. In addition, we invite master teachers from the school district to campus to conduct workshops or seminars with the undergraduates to be used in conjunction with the previously mentioned field experiences. These activities provide prospective teachers with hands-on work with the curriculum that they are able to apply immediately to middle school students.

Newly designed courses and changes made in the original courses

There are two new courses that are of particular importance to the program. These two courses are specifically developed for the middle school majors. *Learning Mathematics with Technology* is a beginning technology course in mathematics education in which preservice teachers explore mathematics through computer applications and other technology resources. They learn new mathematics while simultaneously using computers, calculators, the Internet, and CD-ROM/video disk technologies to pursue their study of mathematical ideas. It was taught during Spring 1996 and proved highly successful.

The second course forming a cornerstone of the program is a "mathematical reasoning and proof" course. It is placed early in the program so that preservice teachers can acquire an understanding of the nature of mathematics, mathematical thought, and mathematical reasoning. Incorporated into this course is a support mechanism built on the Uri Treisman Model (Culter, 1982; Treisman, 1990) for enhancing achievement in mathematics. The teacher is a mathematician who is very interested in mathematics education research, particularly in student belief systems and how learners develop patterns of mathematical reasoning. This teacher designed and field-tested the course in consultation with us. It has been shown to be effective in the eyes of the middle grades majors.

In addition to these two newly designed courses, we are working with the Statistics Department to develop a probability and statistics course appropriate for preservice middle school mathematics teachers. At the same time, the two departments are developing a *statistics lab* to support experimentation and technology utilization in this course.

A middle school methods course is also currently being developed. It focuses on the principal components outlined by the MIDDLE MATH Project but with substantial integration of technology and field experience. Mathematical modeling, problem solving, and teaching mathematical reasoning will be key ingredients. Additionally, authentic and naturalistic assessment will be developed within and modeled in this course. The field experiences in the PAC program, described earlier, and a full semester of internship in a public middle school provide ample opportunities for our majors to practice what they learned.

Other features of our middle school program

In order for our mathematics faculty to be capable of (and comfortable with) using technology in their teaching, we have invited a series of nationally known speakers to campus to offer hands-on workshops for the effective use of technology. For example, three workshops were recently organized to show how to use Maple and the TI-83 in teaching calculus. The workshop series will continue for the next two or three years, and in the near future we plan to invite an expert statistician to campus to talk about technology-based reform statistics curricula.

In addition, we have organized a study group for undergraduate students. The members of this group meet every Friday for five to seven hours. The group's activities include studying innovative curricular materials, exploring effective ways of using technology, and interacting with master teachers for teaching insight. Smaller study groups for individual courses are also encouraged, providing extra opportunities for those with learning difficulties to ask questions and seek more help.

To help preservice teachers with learning difficulties, we have established tutoring systems. First, mathematically strong preservice teachers are identified in mathematics content and methods courses (as well as the newly designed technology course). Then, these preservice teachers tutor weaker undergraduate students ten to twenty hours per week. There are some scheduled time periods for tutoring, or any individual who needs tutoring can contact one of the tutors and set up a specific time and location for meetings.

Issues and Challenges

The changes mentioned previously reflect what we have learned from the MIDDLE MATH Project—specifically, the emphases on appropriate math content, appropriate pedagogy, field experience, and use of technology. The content of course work must have both appropriate depth and appropriate breadth, and it should emphasize problem solving and deeper mathematical reasoning. The pedagogy should feature more active and responsible engagement on the part of pre-service teachers, and the field experience component will continue to develop in depth and breadth. These three aspects are all related to creating a newly designed technology course along with its impact on preservice teachers' learning and teaching of mathematics, and the undergraduates' experience with the PAC program, etc.

Having an opportunity to see several middle grades mathematics curricula projects that had been funded by NSF during the first MIDDLE MATH conference helped us in choosing some of the on-campus school curricula. This experience also contributed to our decision to have undergraduates work with these curricula in conjunction with master teachers from the school system.

Even though our decision to have a substantially technology-based degree program was already forming, the experiences in the MIDDLE MATH Project bolstered our resolve to stick with and increase the emphasis on the technology component. This was partly a reaction to the under-utilization of technology we perceived in other programs as well as to the success we were already having with technology courses. In addition, our philosophy evolved from the extremely positive reactions of middle school students to the technological environment in which they were immersed and the success they were having in understanding sophisticated mathematical concepts using technology.

Finally, the MIDDLE MATH Project probably accelerated the rate of development of the program's components as we developed and tested individual courses before debuting the program as a whole. The discussions, questions, and interactions inherently involved in the creative process helped us to move along more rapidly.

We have discussed in detail two essential elements of the Florida International University Middle School Mathematics Program (which is still under development). Our program has a substantial technology base, as well as varied and extensive practical and field experiences as a second major feature. The program utilizes technology to bring the "Rule of Four" into a conceptual and reasoning-based implementation. The Rule of Four states that, wherever possible, topics should be presented graphically, numerically, analytically, and verbally. Although popular literature tends to trivialize the Rule of Four by suggesting that it simply implies algebraic, arithmetic, pictorial, and verbal forms of mathematics should be used, the original sources and their illustrations all suggest a much more mathematically accurate evocation of the four components and their interrelations which should occur concurrently in a lesson. This means that fitting curves to data, constructing multiple, dynamic, linked representations of important concepts, and becoming able to read even the more complex and subtle mathematical symbols (as well as oral and written communication) are to be frequently done in the mathematics classroom. This is surely an important area of application in the preparation of middle school teachers of mathematics, and it should also be noted that our program focuses on the reasoning involved, specifically the proportional and visual reasoning of middle school students, as emphasized by the NCTM *Standards*.

Because of the PAC program, we make sure that *Standards*-based instruction is modeled and studied in middle school classrooms, as well as practiced with middle school students. Throughout the program, undergraduates experience at least two levels of preparation. They learn the theory involved in the mathematical content, *Standards*-based pedagogy, and psychology components—with due attention paid to the learning styles and psychology of middle school students living in an urban setting. These academic emphases go back to the research-base on which the reform was founded and to the research that supports such significant ideas as the use of the Rule of Four. Accordingly, preservice teachers see and practice modeling these techniques in varied practical settings. Emphasizing the parallels and the interaction between theory and practice is essential, particularly for teachers preparing to teach in complex, multicultural, urban environments.

In this paper, we have not taken the time or space to deal with the full range of the program, nor have we fully described the many barriers to the program's complete success. We welcome any communication and discussion of middle school programs, for such communication can help us expand on successes and overcome difficulties in the future.

References

Culter, K. (1982). *Helping minority students to excel in university-level mathematics and science courses—The professional development program at the University of California, Berkeley.* Prepared for the United States Department of Education, National Commission for Excellence in Education.

Treisman, P. (1990). A study of the mathematics performance of black students in college. *Mathematicians and Education Reform: Proceedings of the July 5-7, 1988 Workshop, University of California, Berkeley.* Providence, RI: American Mathematical Society.

Lees-McRae College

Ita Kilbride
Lees-McRae College

The Institution

Lees-McRae College is a small four-year, private liberal arts institution
established in 1900. Affiliated with the Presbyterian Church, it is
situated in the Blue Ridge Mountains of western North Carolina in the
town of Banner Elk, which has a population of approximately 1,200.

The Program

The middle school mathematics education program is separate from the
elementary (K–6) program. An education professor with mathematics
licensure, a mathematics educator, and a mathematician teach in the
program and supervise students during their student teaching experience. Teachers from a professional development school located beside
the college campus often help teach methods classes to student teachers.

The 37-semester hours program in the Education Department includes a
six-credit sequence — *Organization and Philosophy of Middle School*,
which includes a field experience, and *Materials and Methods for
Teaching Grades 6–9*. Within the Mathematics Department, students
take 15-23 semester hours, depending upon the courses they have taken
in their General Education Core. Courses that satisfy mathematics and
general education requirements include *Pascal, College Mathematics
with Applications, Calculus I,* and *Statistics.* Students must also take
*Finite Methods, Plane and Solid Analytical Geometry, Calculus I Lab,
Advanced Probability and Statistics,* and *Linear Algebra.*

The Accomplishments

In the introduction to *A Call for Change: Recommendations for the
Mathematical Preparation of Teachers of Mathematics* (Leitzel, 1991),
the author describes a vision of the ideal mathematics teacher for the
1990s and beyond. Among other attributes, such a teacher must —

- "Understand how mathematics permeates our lives and how the various threads within mathematics are interwoven.

- Naturally and routinely use technology in the learning, teaching and doing of mathematics.

- Possess knowledge and have an understanding of mathematics that is considerably deeper than that required for the school mathematics he/she will teach." (p. xiii)

The previously mentioned attributes were very much in my mind when I reviewed the middle grades teacher preparation program at my own institution (as part of my preparation for the follow-up MIDDLE MATH Conference in the summer of 1996). Further, the mission statement of Lees-McRae College stresses service to the rural Appalachian community in which it is located. The teacher education program at the college (including the middle school mathematics program) strives to increase the quality of offerings to preservice and inservice teachers in this rural part of western North Carolina.

However, a re-examination of the middle grades mathematics program revealed three challenges to the vision described above.

1. It is difficult to see how mathematics runs throughout our everyday lives in a multicultural society and to include multicultural experiences in math education classes due to the isolated nature of the college's geographic location.

2. Calculators and computers in math education classes were considered as an *add-on* feature rather than being infused throughout the program. There was no consistency with calculator and computer use in mathematics courses.

3. It is difficult to believe that students have a deep understanding of mathematics when mathematics education students themselves have difficulty reaching the required passing score on the PRAXIS II examination.

In order to address these three challenges, several programmatic changes were made. To increase multicultural awareness in the middle grades mathematics program, it was necessary to make innovative and creative use of limited resources. First, it was decided to utilize the cultural heritage of faculty members and other faculty visiting the region. Faculty members from China, Africa, Ireland, and Great Britain shared the educational practices from their homelands with the pre-service teachers. A Fulbright Exchange Teacher from Leister, England (who at the time taught in Burke County, North Carolina) visited the college and discussed multicultural issues in education. Also, it is relatively easy to travel the world electronically to see issues and challenges faced by teachers far removed from the rural setting in western North Carolina. Students used the Internet to visit schools in continental Europe, Singapore, and my home country, Northern Ireland, to learn about a variety of different math programs (see Appendix at the end of this chapter for details).

More traditional means of addressing the challenge of diversity were also included in the program. Students visited inner-city urban school settings, including a technology magnet school, Cook Middle School in Winston Salem. They studied the social, emotional, cultural and educational influences on math education curriculum, and efforts were made to increase recruitment and retention of multicultural students by college faculty and staff. In addition, we tried to maximize the influences of Appalachian culture and heritage on education.

To address the second challenge, attempts were required to infuse technology into mathematics and mathematics education courses. A primary goal of the middle grades math education program is to understand the role of computers and calculators in modern society and to develop competence in using these tools for a variety of problem-solving applications. Therefore, two computer courses will be required in the middle grades mathematics program: *Pascal* and *Calculus Lab I*. Computer and calculator instruction will be stressed not only in methods classes but also in required education courses: *Introduction to Computing* and *Computer and Media Applications in Education*.

The Mathematics Department is attempting to integrate graphing calculators and computers in all appropriate courses. Calculators are used to simplify expressions involving rational expressions and radicals, to perform calculations with logarithms, and to graph various functions and relations. Calculators and computers are often used to analyze problems whose solutions are well beyond conventional pencil and paper techniques. Demonstration examples are taken from the fields of business, physics, statistics, and social sciences.

An additional two texts, which promote the integration of computers and calculators, will be required in middle grades methods courses:

Integrating computers in your classroom: Middle and Secondary Math by Dublin, Pressman, Johnson, and Mawn (1994); and *Mathematics: A Good Beginning* by Troutman and Lichtenberg (1995).

In Fall 1995, the Mathematics Department established its own mathematics/science computer lab with the purchase of five new computers and a printer. The software packages, *Derive*, *Mathematica*, *Maple*, *Minitab*, and *Geometer's Sketchpad,* were purchased for the lab. The computer lab itself is used in the lab portions of the *Calculus I* and *Calculus II* courses, as well as in other courses in the curriculum. This increased emphasis on technology will further prepare the graduating teacher for informed and competent service to middle grades education.

Our attempts to raise scores on the PRAXIS examination may have implications for other teacher education programs. In January 1996, the Education Division from Lees-McRae attended an information-gathering meeting sponsored by Educational Testing Service and Public Schools of North Carolina at Winston Salem State University. The meeting, which described the national emphases and trends in the exam, stressed the profound difficulties with the low test results. There is small comfort in knowing that our math education students are not alone in having difficulty with this exam.

Specialty-area professors, including mathematics, volunteered to sit for the PRAXIS examinations in the spring of 1996. They subsequently met with education faculty to discuss their findings so that changes could be made in course content as well as the mathematics education programs.

The mathematics PRAXIS examination is composed of a content portion and a pedagogy portion. The content portion of the examination contained 50 multiple-choice questions. All material on the exam is covered in the required mathematics classes in the college, including—

- 4–5 questions from geometry

- 4–5 questions from probability

- 8–10 questions from *Calculus I* and one question from *Calculus II*

- 6–8 questions from *Linear Algebra* and one from *Modern Algebra*

- 4–5 questions from computer programming, mostly involving interpreting the given programs

- 4–6 questions involving proofs and mathematical reasoning

- 8–10 questions from *Algebra I* and *II*.

The pedagogy portion required students to prepare three lessons in an exam period of one hour. They had to show how they would introduce each lesson, describe its design, and give appropriate examples and problems. In the time given, they also had to discuss assessment of the lesson and indicate what manipulatives and software would be useful for the lesson.

At Lees-McRae, students completed a series of PRAXIS preparation workshops. A sample test was administered and sample test items were analyzed for content and rationale. For the future, we decided pre-service teachers will include the five stages of introduction, design, examples of problems, assessment of the lesson, and appropriate manipulatives and software in their six-point lesson plan preparation. We will have them practice lesson design under tight time constraints, and we will include topics similar to those on the PRAXIS exam in their unit plans and lesson plans in middle grades math methods class. We will also include mathematical applications of software in the required twelve software-evaluation exercises that students complete in their methods class.

Issues and Challenges

Since my participation in the follow up MIDDLE MATH Conference in the Summer of 1996, the mathematics professors in the Division of Science and Mathematics have developed and introduced a new course in mathematics, *Analysis of Basic Algebra*, specifically designed to prepare future mathematics teachers for the mathematics section of the Praxis II: Subject Assessments. Participation in the MIDDLE MATH Project has heightened my awareness and the awareness of my colleagues at Lees-McRae to the responsibilities and challenges inherent in the preparation of teachers of mathematics.

References

Dublin, P., Pressman, H., Johnson, W., & Mawn, B. A. (1994). *Integrating computers in your classroom: Middle and secondary math.* New York: Harper Collins.

Leitzel, J. R. C. (Ed.). (1991). *A call for change: Recommendations for the mathematical preparation of teachers.* Washington, DC: Mathematical Association of America (MAA).

Troutman, A. P., & Lichtenberg, B. K. (1995). *Mathematics: A good beginning: Strategies for teaching children.* (5th Ed.) Pacific Grove, CA: Brooks/Cole Publishing Co.

Appendix: Web Addresses

J. P. Fay, European Studies Office, 153 Bangor Road, Holywood, Co. Down BT18 OEU, Northern Ireland. Email Address: esp@iol.ie or jpfay@delphi.com. Web Address: http://www.infm.ulst.ac.uk/~esp/resource/nisites.html or http://www.iol.ie/esp/resource/nisites.html.

Knockloughrim Primary School. Web Address: http://www.neelb02.demon.co.uk/knockrim/index.html

Bellaghy Primary School. Web Address: http://www.rmplc.co.uk/eduweb/sites/neelbetc/schools/ bellaghy/index.html

Tamlaght O'Crilly, Drumard Primary School. Web Address: http://www.drumard.demon.co.uk/

Mercer University

Mercer University

The Institution

Mercer University was founded in 1833. The second-largest Baptist-affiliated institution in the world, it is the only independent university of its size in the country that combines programs in liberal arts, business, engineering, education, medicine, pharmacy, theology, and law. Mercer's main campus is located in Macon, a city of 120,000 residents that serves as the hub of Central Georgia. With five off-campus centers, Mercer's educational programs reach virtually every corner of the state.

In 1995, teacher education programs previously offered through Mercer's College of Liberal Arts and academic programs once offered through its University College were consolidated to form the School of Education. Currently 1,700 students strong, the School of Education is one of the largest schools at Mercer.

The Program

The middle grades mathematics teacher preparation program at Mercer University has historically consisted of a mathematics methods course offered through the School of Education and 25 credit hours of mathematics courses offered by the College of Liberal Arts. Four mathematics courses (five credits each) are required for a middle grades mathematics concentration. Students choose from *Applied Mathematics*, *Consumer Mathematics*, or *College Algebra*; *Business Statistical Methods* or *Data Analysis for the Social Sciences*; *Calculus I* and *II*; *College Geometry*; and *Special Topics in Mathematics*. All middle grades preservice teachers, regardless of their concentration, must take *Mathematical Problem Solving* and *The Modern Approach to Pre-Calculus*. However, any of these courses are only as effective as the mathematics instructors teaching them.

In the past, little communication occurred between the Colleges of Education and Liberal Arts. As a result, a perception developed that "real" mathematics was presented through lecturing and that mathematics methods were akin to playing games with little connection to true mathematics. Using knowledge gained from the MIDDLE MATH Conferences, the middle grades mathematics teacher preparation program at Mercer is in transition. Although many changes are being implemented and characterize this transition, this case study will analyze the development and implementation of one particular course: *Mathematical Problem Solving*.

The Accomplishments

Mercer University's School of Education had initiated the use of a standardized exam (a pre-professional skills test known as PRAXIS I) to quantify competency in mathematics and English skills. The first time the exam was administered, 83% of the students in the teacher preparation program failed — even with a cut-off score at the 13th percentile level. To address these inadequacies, the Mathematics Education Department first required tutoring before allowing students to retake the exam. However, tutoring did not increase the number of students passing the mathematics portion of the PRAXIS I exam. At this point, the mathematics education faculty initiated discussions on implementing a course to improve self-efficacy relative to mathematical problem solving.

There already existed a mathematical problem-solving class (taught in the Mathematics Education Department) which required students to use a very structured problem-solving strategy based upon the four steps outlined by Polya. Students covered the typical pre-algebra content and were given problems to solve in each pre-algebra area. Inspired by the homework reading from the MIDDLE MATH Project, we realized that the structure of this class not only was ineffective but also failed to reflect the best practices of how people learn mathematics. Consequently, we decided to overhaul the course. Using knowledge from the theory of constructivism, the NCTM *Standards*, and the text *On the Shoulders of Giants* (Steen, 1990), the faculty designed a very nontraditional course. The uniqueness of each student's learning style, cultural experiences, and ways of creating mathematics was celebrated. The course recognized that there are many ways to make meaning of mathematics and valued different ways of thinking.

First, the new mathematical problem-solving course requires that each student be responsible for his/her own mathematical learning. The pedagogy centers around the use of the NCTM *Standards* and the theory of constructivism. Two rules dominate the class: (1) No one can tell another how to work a problem, and (2) No one can tell another whether the answer is right or wrong. Initially, these rules created much confusion. There were phone calls and visits to the dean with the complaint, "I paid $500 for this course and that teacher will not tell me how to work the problem." Fortunately, the dean was well versed on how constructivism worked and supported both the structure of the class and the instructor. Trying to be proactive, several articles on the application and effectiveness of the theory of constructivism in the classroom were used to stimulate discussion concerning the pedagogy used in the class. One of these articles (Clements & Battista, 1990) is now required reading and is used to reach consensus at the beginning of the course for understanding the structure of the class.

The content of the course is still based in pre-algebra mathematical concepts, but now is organized using topics such as Patterns and Connections. A topic will appear more than once during the semester; as the topics cycle through, they will build on themselves. Each class session begins with mathematical problems to be solved. These problems are designed so they can be solved from more than one perspective. An example of one such open-to-interpretation problem is as follows:

> Give each group of students some string, beads and various objects totaling to 26. Ask them to make a four-stranded necklace.

The group must agree on the problem-solving strategy and the finished product. The discussion, compromise, and interpretation process is always rich. Some students use division by tearing the beads, resulting in 6 1/2 beads on each strand. Others create uneven numbered strands using an artistic approach of estimating and evaluating the aesthetic value of the finished product. Others use division and make four necklaces of six beads each and use the extras for earrings. My personal favorite is a necklace of four tiers with decreasing numbers of beads on each tier. It was modeled after an African necklace one of the students owned. All the preservice teachers and their instructors were amazed at the many "correct" interpretations and solutions constructed from a very simple mathematics problem.

Using mathematical problems similar to the previous illustration, the students work in groups explaining their mathematical thought processes as they attempt to solve each problem. The students and instructors try to follow the rules of not distributing information. Instead, the students search in various sources for pictures, manipulatives, or teaching strategies to help each other. The most difficult part of this course for the instructors has been not to tell the students how to work the problems. The nature and experiences of most teachers puts them in the role of helping the students. To have students be frustrated and not be able to assist them in the manner to which they are accustomed is very trying. To develop appropriate strategies to ask good questions in order to stimulate student thinking and initiative is hard work, but it is rewarding. By mid-term, most of the students are thinking on their own and not dependent on the instructor to "teach them."

Issues and Challenges

The initial goal of this project was to improve the students' mathematical self-efficacy relative to problem solving. The anecdotal evidence for this is overwhelming—students who are passive and waiting for someone to make mathematics meaningful for them become quite vocal in the group discussions. Most students' evaluations report that they "feel more confident about being able to solve problems." An unanticipated result of this approach has been that preservice teachers realize others may process and interpret mathematics differently, but still validly. It is expected that this knowledge should improve their ability to teach all kinds of students. Preservice teachers will have a rich supply of mathematical strategies, and experience communicating and reasoning mathematically from their group work in this class that will allow them to interact with their students more effectively.

The course is now being further modified. To bridge the gulf with the Mathematics Department, mathematicians will be invited to help in the development of the problems for the course and to attend and team teach the classes. It is expected that as these mathematicians are convinced that real mathematics can be learned without lecturing, this class will be eventually turned over to the mathematics faculty to teach.

The pilot project of this course in 1996 received outstanding evaluations from the senior class. The students' standardized test scores improved from a 83% failure rate on the PRAXIS I to a 36% failure rate. Journal reflections showed the students gained deeper understanding of how others can think differently and yet have no less problem-solving ability than themselves. The students were excited about the power they felt this gave them in helping their future middle grades students.

References

Clements, D. H., & Battista, M. T. (1990). Constructivist learning and teaching. *Arithmetic Teacher*, *38*(1), 34-35.

National Council of Teachers of Mathematics. (1989). *Curriculum and evaluation standards for school mathematics*. Reston, VA: Author.

Steen, L. A. (Ed.). (1990). *On the shoulders of giants: New approaches to numeracy*. Washington, DC: National Academy Press.

The Ohio State University

Douglas T. Owens
The Ohio State University

The Institution

The Ohio State University is one of the largest public universities in the country. As a comprehensive teaching and research university, it is Ohio's premier institution of higher education. The Integrated Master of Education Program in Mathematics, Science, and Technology Education described in this case study is situated in the College of Education's School of Teaching and Learning. This structure of the College of Education is in its first year of operation at the time of writing.

The Program

In about 1990, the College of Education made a decision to adopt the Holmes Group Plan of teacher education — so students first earn a Bachelor's Degree in the subject area they want to teach and then complete their teacher education course work at the graduate level. At Ohio State, a Master of Education program is designed to be completed over five quarters, beginning in the summer and finishing the next summer. The new integrated Mathematics, Science, and Technology Education program was instituted in 1996-97. The program leads to high school certification — but we designed an endorsement in mathematics and science for middle school teachers to be attached to that certification. Although secondary certification is dependent upon the major subject, middle school certification is required in two subjects, and technology education (formerly industrial education) is a K–12 certification.

Students with an undergraduate mathematics major would be certifiable in middle school mathematics and would take 35-quarter hours of science courses in life sciences, physical sciences, and earth sciences. Those with science majors might take additional mathematics courses to meet the state 30 quarter hour minimum. Among the courses accepted as suitable are *Mathematics for Elementary School Teachers*, *Finite Mathematics*, (non-calculus) *Statistics*, *College Algebra*, *Elementary Functions*, and a variety of courses in *Calculus*. Additionally, higher-level courses may be counted for credit rather than lower level

courses. Because, in Ohio, one can teach in middle school with certification in either elementary school or a high school subject area, the
new program offers an attractive alternative to the traditional route to
middle school teaching (which has been through a program in early and
middle childhood education).

An overriding theme or philosophy of the new program is that the
school subjects of math, science, and technology should be integrated
to some extent. In virtually all courses, we expect integration of concepts of math, science, and technology. We also expect the participants
in the program to have some background in both math and science to
make this integration possible. A second philosophy of the program is
that knowledge is constructed by the knower. A belief in constructivism
results in the theme of learning by hands-on activities and projects. A
goal of the program (and an evolving reality) is to have hands-on class
activities and assignments in integrating concepts in mathematics and
science supported by computer or graphing calculator technology. The
program is self-contained in that all aspects of the program are deliv-
ered by faculty and graduate teaching assistants in mathematics, science, and technology education. This practice is a statement of a strong
belief in subject matter as a driving force in teacher education. Another
practice that contributes an important philosophical ideal is that of team
teaching. Each course in the program in the first year of operation (with
the exception of the mathematics methods course) has been taught by a
team of instructors from at least two of the three fields.

Within the program, there are a series of three integrated pedagogy
courses—*Integrated Pedagogy I, II,* and *III*. The introductory course
focuses on the major curriculum goals and conceptions of the three
fields of mathematics, science, and technology education and relates
these goals to more general trends in the history and philosophy of
education in this country. The second is a course in curriculum, instruction, and assessment with attention to integrating subject matter in
middle and secondary schools. The third course investigates issues of
diversity and equity with special attention to practices in classroom
management and assessment.

In addition to the integrated pedagogy courses, there is a specialty
methods course in each of the three areas of mathematics, science, and
technology education. The methods course in mathematics teaching has
a strong component in the use of technology to support the teaching of
mathematics courses and topics in middle and high school.

Because there were formerly no courses at the university on the integration of content, a sequence of three integrated content courses was
created. The goals of the courses are to focus on problems and activities
that integrate concepts and processes across the various areas of mathematics (arithmetic, algebra, geometry, measurement, and data analysis),
science (life science, earth science, physics, and chemistry), and technology (manufacturing, construction, communication, and power).
Problems are presented as inquiry activities accessible at the middle
school level while being open-ended enough to be challenging at the
high school level. Consideration has to be given to adapt the activities
to accommodate backgrounds in mathematics, science, and technology
and to maintain the integrity of each discipline while integrating across
disciplines.

The new program also includes three courses (*Fundamental Concepts
of Mathematics*, *Psychology of Learning Mathematics and Science*, and
a research methods course) designed to address the fundamental ideas
in the disciplines of mathematics, science, and technology. An overarching goal of *Fundamental Concepts of Mathematics* is the development of the nature of mathematics as a discipline. The course is designed to help students understand how some of the key concepts and
processes of mathematics play out in the middle and high school
mathematics curriculum. Many of the activities are taught by the
inquiry approach to illustrate how students may engage in a variety of
topics at a variety of levels.

Psychology of Learning Mathematics and Science combines two
formerly separate courses into an integrated course team-taught by a
mathematics educator and a science educator. The course deals with the
psychological underpinnings of mathematics and science education and
references the literature that is typically cited by mathematics and
science education researchers and curriculum developers.

The program contains a research methods course that provides an
overview of both quantitative and qualitative methods of inquiry. The
course serves as a foundation for framing problems and conducting
investigations especially appropriate for teachers' action research. This
activity ties in with the action research project which students must
complete before they finish the program. Program participants are
expected to consult with their advisors and cooperating teachers in the
field to negotiate a suitable project. Data is typically collected during
the student teaching experience—the only opportunity that allows

protracted involvement in the field. The final project report is submitted to each student's masters' committee for approval.

Finally, The Ohio State University's five-quarter Master's degree program has a field component each of the three quarters during which middle and high schools are in session. During the Autumn Quarter, interns are in schools four mornings per week. Activities include observing, tutoring, marking, assisting, and teaching some lessons. On the fifth morning, interns are on campus for a clinical experience course. The clinical experience course includes reflection on and discussion of field experiences, peer teaching, and micro teaching. Interns are in the field five mornings per week during Winter Quarter. They are assigned to teach one class for five days per week for seven weeks. They are expected to teach some lessons using a variety of instructional methods, technology, alternative assessments, and integration of mathematics with another subject. They are also expected to spend one weekend assisting children at the Center of Science and Industry. Interns are encouraged to participate in extra-curricular activities, parent conferences, teachers' meetings, and observing teachers and fellow interns. They write a weekly reflective journal and build a professional portfolio. During the Spring Quarter, interns are involved in student teaching all day for ten weeks, and they are expected to build to and maintain a full teaching schedule for six weeks. Lesson plans are to be completed a week in advance for peer and supervisor critique. Supervision is conducted at least weekly and at least three conferences are held involving supervisor, intern, and mentor teacher. Supervision is primarily conducted by Graduate Teaching Associates who are doctoral students under faculty oversight.

The Accomplishments

The Integrated Master of Education Program in Mathematics, Science, and Technology Education was begun in the summer of 1996. A series of meetings over the previous academic year had led to decisions and descriptions permitting the establishment of the program. Many other decisions were made ad hoc as we went along — all but one of the courses described above are new, including the psychological foundations and the mathematics methods course. Although a M.Ed. program in mathematics education had been in operation for five years, the integrated program (as well as working with faculty and graduate teaching associates from science and technology education) was new to us. The beginning of the new program was concurrent with a restructuring of the College of Education into three schools and developing and learning new procedures and ways of working in new relationships.

At this writing, official course approvals are in progress for two of the integrated sequences detailed above (the pedagogy sequence and the fundamental ideas sequence). The content courses are still in development.

Issues and Challenges

Pressures abound against the self-contained nature of the program. Research methods often seem to be an arena for jurisdiction claims, and regularization of the research course component has yet to be tackled. In a recent meeting, conflict resurfaced about using service courses in other departments for components of sociological foundations and issues of equity and inclusion.

Further, in the state of Ohio, legislation has been passed for new teacher licensure standards. The new standards require eighteen quarter hours of reading courses for middle school certification and this requirement puts our new middle school endorsement in mathematics and science education in jeopardy. We will need to work over the next two years with faculty in Reading Education to develop a new program. The strong concentration required in reading may well force reading to become one of the two teaching areas for many teachers instead of the expected science-mathematics or technology-mathematics combinations.

Saint Xavier University

Susan Beal and Eileen Quinn Knight
Saint Xavier University

The Institution

Saint Xavier University is a coeducational, progressive, urban, and international university founded in the rich tradition of the Catholic faith. Located on the southwest side of Chicago, it serves over 4,000 students with nationally respected programs in nursing, business, education, and the liberal arts and sciences. Saint Xavier's excellent facilities, small class size, and dedicated faculty are just a part of the quality education that has been offered at the university for 150 years.

Although mathematics education classes at Saint Xavier tend to have under 25 students, the number of faculty able to teach these classes is also small. Offering new classes on a rotational basis allows us to teach a variety of classes which meet the requirements of the State of Illinois while still meeting the needs of our undergraduates.

The Program

In lieu of a master's program for middle school mathematics (which was recently suspended), it was decided that we would enrich our course offerings at the undergraduate level for prospective middle school teachers. Four new courses were proposed: *Number Theory for Teachers*, *Probability and Statistics for Teachers*, *History of Mathematics for Teachers*, and *Geometry for Teachers*. These courses were approved by the Faculty Senate and we have been teaching one of these courses each semester starting in the Fall of 1996.

In addition to the four new courses, the Mathematics Department will continue to teach *Mathematics and Methods for Middle and Junior High School*, an integrated mathematics and methods course. The prerequisites for this course (and the four new courses) are the two mathematics content courses: *Foundations of Mathematics I* and *II*. The two content courses were, until this past year, integrated mathematics and methods courses. The methods course is now offered by the School of Education—as required by the state— and it will be described in detail.

To receive an endorsement in middle school mathematics from the
State of Illinois, a student must have three hours in middle/junior high
school mathematics methods and fifteen hours in mathematics satisfy-
ing four of the following areas: elementary mathematics and methods,
geometry, number theory or abstract algebra, linear algebra, calculus,
probability and statistics, and history of mathematics. The students at
St. Xavier have four options for meeting these qualifications. They can
receive a middle school mathematics endorsement as a mathematics
education major in the Department of Mathematics and Computer
Science, an elementary education major with a concentration in mathematics, a minor in mathematics, or a minor in mathematics education.

With the addition of the four new courses to the mathematics education
curriculum, both the mathematics education major and minor are
redesigned. A student majoring in mathematics education takes the
following courses in the department of mathematics: *Calculus I* and *II*
(8 hours), *Introduction to Discrete Mathematics* (4 hours), *Probability
and Statistics for Teachers* (3 hours), *Linear Algebra* (4 hours), *Modern
Geometry* (3 hours), *Abstract Algebra* (3 hours), *Middle School
Mathematics and Methods* (3 hours), a 200-level computer science
course (4 hours), and *Senior Seminar* (1 hour). The *Foundations of
Mathematics I* and *II* are prerequisite courses adding six additional
hours to the 33 hours required for the major. Recommended electives
are *Number Theory for Teachers* and *History of Mathematics for
Teachers*. The student must also satisfy the education courses required
by the state and the School of Education. This major was designed for
students interested in teaching at the middle school level with a specialty in mathematics or being a consultant or coordinator of mathematics for a school or school district.

Elementary education majors may also minor in Mathematics Education. The following sequence of courses is suggested for them: *Introduction to Discrete Mathematics* (4 hours), *College Algebra* or *Calculus I* (3 or 4 hours), *Foundations of Mathematics I* and *II* (6 hours),
Middle School Mathematics and Methods (3 hours), a 200-level computer science course (4 hours), and any two of the four new courses (6
hours). To have a mathematics concentration, students must take 18
hours of mathematics beyond the six hours required by the State of
Illinois (that is *Foundations of Mathematics I* and *II*) with 9 of these
hours at or above junior level courses.

It should be noted that, for at least *Foundations of Mathematics I* and *II*
and *Middle School Mathematics and Methods*, students investigate

various areas of mathematics while working in small cooperative groups in a laboratory setting, using manipulatives, scientific calculators, and computers when appropriate. Students are required to keep journals to reflect upon the way(s) they are (re)-learning mathematics.

The Accomplishments

In this section we describe *Math and Cognition for the Middle School*—a two-course sequence in elementary and middle school mathematics in the College of Education (which Knight designed and implemented while completing her dissertation). Our emphasis will be on three aspects of the second semester course dealing with middle school math preparation: the journal entries, the scrapbook, and the exit interviews. Students in the course were from our regular program in teacher preparation. A pre-test was given at the beginning of each semester to assess basic computational, problem solving, and pedagogical skills. It was found that students in the mathematics methods course had paltry knowledge of middle school and elementary mathematics.

The course emphasized constructivist and collaborative approaches to teaching and learning. Attention was also paid to emotional issues. The course focused on two issues of reform recommended by the National Council of Teachers of Mathematics (1989). First, there was an emphasis on the difference between memorizing the steps of a procedure to get a correct answer and understanding both mathematical symbols and the logic of a mathematical solution. In particular, the mathematics of fractions and decimals was stressed. Second, there was an emphasis on encouraging students to participate socially. They were encouraged to share both their efforts to achieve understanding and their feelings. This sharing was a way for the students to become active participants in the course, and helped them realize how much the course was comprised of their own effort and work. Students also realized that others experienced similar feelings and rejected the idea that any individual was less adequate than another. Classes met as a whole group with the teacher and in small groups with students working by themselves asking the teacher to assist them when needed. Additionally, students spent significant time solving or creating their own word problems.

Journals

Each student created a journal in which he or she wrote a reflective entry after each class. The journals concerned themselves with cognitive, social, and affective variables influencing the classroom. The students were encouraged to express themselves freely about issues they faced in doing mathematics, such as the satisfaction or pleasure they felt regarding class activities, what they had learned, personal concerns they experienced, changes they felt as they went through the class, and other educational issues. Students wrote such things as:

> *Learning a few things well is ultimately better than learning many things incorrectly or insufficiently. We often move on to other issues too rapidly without really constructing anything. We need to take time to understand.*

Another student commented she had resorted to memorization as a strategy for coping with the need to learn mathematics that had little meaning for her. She also wrote about the consequences of this way of coping.

> *My reason for taking this class is that I don't want a child to go through what I went through in math class. I had a high level of math anxiety. I remember my class was working on square roots. I did not understand it at all. When it came time for the test, I pretended to be sick and went to the office to lie down. I want to pass something better on to the students I teach.*

> *I never viewed math as fun. It was something that you learned how to do. You memorized, you learned the feature, and then over the summer you forgot the procedures.*

> *My past math experience really screwed up my brain. That was all rote procedures and memorization. Everything was rote and you were completely lost if you didn't hold onto all the information.*

A third student indicated a problem with the quantity of her homework:

> *There was so much emphasis on doing many problems. We have to do twenty-five problems instead of doing four and [trying] to make the connection to real life.*

Sometimes, students indicated that they were ashamed of the way they felt in math class when they were working toward constructing understanding of a problem:

*I can remember hiding the scrap paper because we weren't sup-
posed to use it. We used to do our work on the scrap paper and
then transfer it to the one we were going to hand in.*

These passages confirm the effects of a drill-and-practice method of
mathematics instruction and provide us with an understanding of the
damage done to students through their past experiences with mathematics. Many of the feelings the students expressed about their abilities to
achieve mathematical understanding were negative. After coming to
grips with these negative feelings, preservice teachers were able to
begin to construct mathematics that had meaning for them. Once they
actively participated in building their own mathematical understanding,
the preservice teachers began to prefer to learn mathematics in their
own way.

Scrapbooks

All preservice teachers created a scrapbook that contained the work
they created in the class, either by themselves or in their group and
articles on middle school mathematics that they brought to class. The
preservice teachers collected over 26 articles that they explained to
their colleagues. They also wrote a variety of word problems in their
scrapbooks on the topics we were covering in class. These included a
variety of ways of looking at multi-digit multiplication, proportions,
percents, decimals and fractions, statistical data and probability, measurement, and algebraic word problems. The preservice teachers organized the scrapbook in such a way that it became the textbook they
would use when they went to teach. When they were in the middle
schools, they often shared their scrapbooks with other teachers. These
teachers commented that they learned a great deal from the preservice
teachers by using their scrapbooks and having them explain the concepts and activities contained in them. The students regarded the
construction of the scrapbook as a very positive activity and they
showed a great sense of pride and ownership in their work.

Exit Interviews

At the end of the course, the preservice teachers had an exit interview
to discuss what occurred during the class. They expressed a preference
for learning through understanding that had not existed before. In
response to the teacher's question, "Is there any aspect of the course
that you really liked or that you would point out as something that

needs to be changed?" the preservice teachers replied that doing word problems is probably their biggest accomplishment.

> *I never could do word problems. I didn't know what they were asking or what they wanted. Now I can sit down and do any word problem.*

> *I honestly thought that, in a lot of situations, there was never more than one way to do math (until I got to this class). My grade school teachers forced one way of doing things on me.*

> *I think the good thing about word problems is that you can't do rote memorization. You can't say [while reading a word problem] four times five is twenty. You can't take a word problem and just pump out an answer. You have to understand.*

From the collected data, it seems that preservice teachers achieved a sense of confidence about their ability to work out an understanding of the mathematics in our course. From the pre-test and journal writing, it was clear that, despite their many years of schooling, these preservice teachers seemed to bring to this course a lack of understanding of the middle school math that they would eventually be required to teach. Many of them seemed to have established only in a very limited way the links between mathematical operations and the culture's standard mathematical sign systems and meanings. During the course, students indicated that they were able to construct mathematical understanding, and by the end of the course, they were confident in their mathematical abilities.

Issues and Challenges

It is important to reform our teacher preparation programs so that future teachers will be able to emphasize the building of understanding of mathematics. As preservice teachers in our teacher preparation programs come to prefer and insist on the construction of understanding in mathematics (and in other subjects), they will be able to assist students in building understanding in mathematics, draw up new curricula documents that emphasize understanding, be aware of authentic ways of testing, and motivate our children to want to make sense of mathematics.

It is also important to include mathematics courses that are challenging to the students' understanding of mathematics. The opportunity to revisit middle school mathematics from a different perspective and add depth to students' knowledge of mathematics enables these future teachers to gain the necessary confidence to teach mathematics in a meaningful manner.

Although we have offered *Math and Cognition for the Middle School* as an example, we would like to emphasize that other courses in our program are conducted in an active constructivist environment using small cooperative groups, manipulatives, and technology. The importance of communicating mathematics — listening, speaking, reading and writing — is also emphasized in these classes. Ultimately, it is the blending of what we teach with how we teach it that will impress our students to adopt/adapt a teaching style different from their negative school experiences and add strength to their persona as a middle school mathematics teacher.

References

National Council of Teachers of Mathematics. (1989). *Curriculum and evaluation standards for school mathematics*. Reston, VA: Author.

University of North Carolina at Chapel Hill

Susan N. Friel
University of North Carolina at Chapel Hill

The Institution

The University of North Carolina at Chapel Hill (UNC) is a public university located in the central region of North Carolina. Chartered in 1789, UNC was the first university to be established in the state. UNC is not only the oldest institution in the state, it is the largest and most comprehensive as well. The University serves 24,439 students and offers instruction in over 100 fields.

The University has been educating teachers since it opened its doors to students in 1795, making it one of the oldest teacher education institutions in the country. The School of Education offers programs for the preparation of preschool, elementary, middle, and secondary teachers, as well as other school personnel.

The Program

The Middle Grades Teacher Education Program at UNC is one of North Carolina's model clinical undergraduate programs for teacher education. Graduates of the program earn Bachelor of Arts degrees in Education and are licensed to teach in grades 6–9 in two of the following four subjects: English/language arts, mathematics, science, or social studies. Prospective students apply to the School of Education each spring, and 20 to 25 students are admitted to Middle Grades Education. Each cohort progresses as an interdisciplinary group through carefully sequenced courses over a two-year period.

The senior year internship involves a yearlong placement. During the fall semester, an interdisciplinary team of professors models teaching methods and demonstrates materials used in the content areas of English/language arts, mathematics, science, and social studies. Integrated into this methods course is instruction about how to teach reading and writing in all content areas and using technology for teaching and learning. Students work in their internship sites for the first three days of public school and then two hours per week during the fall semester. Students prepare unit plans in two content areas and will teach one of these units in the spring. Additionally, full-time student teaching in that same class is done by students from early January through the end of April. Most interns teach in one subject, although split placements have been occasionally assigned.

Students may elect to have mathematics as their primary concentration (24 credit hours) or as their secondary concentration (18 credit hours). Mathematics content addressed in both concentrations includes *Calculus of Functions of One Variable* (3 credits) and *Calculus of Functions of Two Variables* (3 credits), *Discrete Mathematics* (3 credits), *Elementary Probability and Statistics* (3 credits), and two new mathematics courses focused on topics related to the middle school mathematics curriculum—*Revisiting Real Numbers and Algebra* (3 credits) and *Topics in Geometry* (3 credits). These two new courses provide opportunities to revisit mathematics topics in ways that allow students to develop a deeper understanding of the subtle ideas and relationships involved in the content of middle school mathematics. In addition, students focusing on mathematics as a primary concentration choose further course work (6 credits) from *Linear Algebra*, *Elementary Number Theory*, *Euclidean and Non-Euclidean Geometries*, the *History of Mathematics*, and *Introduction to Computers*.

The Accomplishments

Central to the preparation for teaching mathematics is the development of a deep understanding of the mathematics of the school curriculum and how it fits within the discipline of mathematics. Too often, it is taken for granted that teachers' knowledge of the content of school mathematics is in place by the time they complete their own K–12 learning experiences. Teachers need opportunities to revisit school mathematics topics in ways that will allow them to develop deeper understandings of the subtle ideas and relationships that are involved between related concepts (NCTM, 1991, p. 134).

In Spring 1995, the decision was made to develop the two new mathematics courses (noted above), focusing on topics related to the middle school mathematics curriculum. In spring 1996, the first course was offered to students. The goals for this course are—

1. To broaden and deepen mathematics understanding in the areas of real numbers and algebraic concepts as topics that are directly related to the middle school curriculum. This includes—

- Developing a practical, concrete sense of real number and algebraic concepts.

- Understanding how the mathematics of real numbers and algebra may be *well* demonstrated (or not *well* demonstrated) through the use of concrete models and real-world applications.

- Representing situations using tables, graphs, verbal rules, and equations, and exploring connections between these representations.

- Understanding what it means to explore solving linear and non-linear equations using concrete, informal, and formal methods.

2. Providing for the development of skills in—and a positive disposition toward—problem solving, mathematical reasoning, mathematical modeling, and the use of technology in the process of doing mathematics. This includes—

- Developing (inductive and deductive) proofs related to the mathematics of real numbers and algebra.

- Using computers and calculators as tools to explore real numbers and algebraic concepts.

- Exploring diverse examples of functions arising from a variety of problem situations and investigating the properties of these functions through appropriate technologies, including graphing utilities and graphing calculators.

The course begins with a focus on problem solving. Students complete a number of explorations related to figurate numbers; topics include the use of inductive reasoning, finite differences, and the application of inductive proof techniques in this applied setting. A focus on deductive reasoning is demonstrated by an exploration of *Attribute Blocks*™, presenting a variety of problem contexts that themselves lead to a more formal discussion of the foundations of set theory. This introduction to set theory helps students clarify the nature of set representations used in the development of number concepts introduced later in the course.

Topics focusing on number theory, mathematical systems, and the real numbers are also included in the course. For example, in their work with number theory, students consider even and odd numbers using a problem taken from the Connected Mathematics Project that involves middle grades students in reasoning about odd and even numbers (Figure 1).

3.3 Reasoning with Odd and Even Numbers

An **even number** is a number that has 2 as a factor. An **odd number** is a number that does not have 2 as a factor. In this problem, you will study patterns involving odd and even numbers. First, you will learn a way of modeling odd and even numbers. Then, you will make conjectures about sums and products of odd and even numbers. A *conjecture* is your best guess about a relationship. You can use the models to justify, or prove, your conjectures.

" AN ODD NUMBER "

Will's friend, Jocelyn, makes models for whole numbers by arranging square tiles in a special pattern. Here are Jocelyn's tile models for the numbers from 1 to 7.

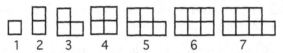

Discuss with your class how the models of even numbers are different from the models of odd numbers. Then describe the models for 50 and 99.

Problem 3.3

Make a conjecture about whether each result below will be even or odd. Then use tile models or some other method to justify your conjecture.

A. The sum of two even numbers

B. The sum of two odd numbers

C. The sum of an odd number and an even number

D. The product of two even numbers

E. The product of two odd numbers

F. The product of an odd number and an even number

Problem 3.3 Follow-Up

1. Is 0 an even number or an odd number? Explain your answer.

2. Without building a tile model, how can you tell whether a sum of numbers—such as 127 + 38—is even or odd?

Figure 1: *Reasoning with Odd and Even Numbers* (Reprinted with permission of the Connected Mathematics Project from Lappan, Fey, Friel, Fitzgerald, & Phillips, 1998, pp. 28-29)

As presented, the problem is designed to provide a way to model even and odd numbers. Middle school students are asked to make conjectures about the properties of odd and even numbers and to build justifications for their conjectures using the models made with tiles.

Many students in this course have had prior formal study in number theory and can reason using both the models designed for middle grades students and the formalizations to which they have been exposed earlier (Figure 2).

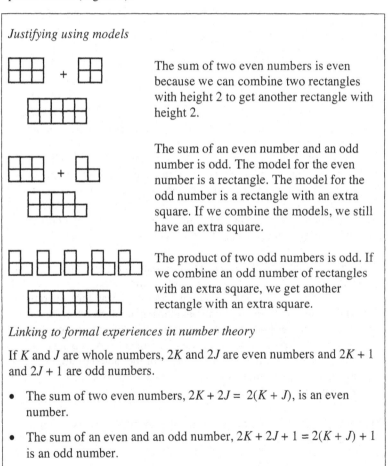

Justifying using models

The sum of two even numbers is even because we can combine two rectangles with height 2 to get another rectangle with height 2.

The sum of an even number and an odd number is odd. The model for the even number is a rectangle. The model for the odd number is a rectangle with an extra square. If we combine the models, we still have an extra square.

The product of two odd numbers is odd. If we combine an odd number of rectangles with an extra square, we get another rectangle with an extra square.

Linking to formal experiences in number theory

If K and J are whole numbers, $2K$ and $2J$ are even numbers and $2K + 1$ and $2J + 1$ are odd numbers.

- The sum of two even numbers, $2K + 2J = 2(K + J)$, is an even number.

- The sum of an even and an odd number, $2K + 2J + 1 = 2(K + J) + 1$ is an odd number.

- The product of two odd numbers, $(2K + 1)(2J + 1) =$ $4KJ + 2K + 2J + 1 = 2(2KJ + K + J) + 1$, is an odd number.

Figure 2: Reasoning about Even and Odd Numbers

Asking students to discuss the connections between their work with the models and their work with the more formal proof structures encourages the development of a deeper understanding of what it means to reason within the context of even and odd numbers.

In the area of mathematical systems, students develop numeration systems that match a given set of constraints (Bassarear, 1997). Building upon this development, they revisit ancient numeration systems and briefly explore numbers represented using different bases. In their work with different number bases, they are asked to consider what makes a number even or odd in bases such as base five or base eight; e.g.,

$$\text{Are } 12_{\text{base five}} \text{ and } 12_{\text{base eight}} \text{ even or odd numbers?}$$

As part of this work, students construct their own concept maps of the real numbers. Concept maps (Bartels, 1995) can help students organize their knowledge about a subject area and evolve from thinking about a big idea (e.g., real numbers) to identifying concepts related to the big idea. For example, Figure 3 shows a partial concept map of the real numbers developed by one of the students during the course. The section of the map shown reflects the connections the student made concerning the operations of addition, subtraction, multiplication, and division (Bartels, 1995) and the relationships of selected properties to these operations. The map provides insights into the student's current thinking about this component of the structure of real numbers.

Another component of the course addresses the question, "How effectively does a concrete model demonstrate the mathematics it is intended to address?" One context involves revisiting integers and the operations of addition and subtraction demonstrated using a colored chip (set) model and a number line (measurement) model. The use of models for integers is extended to consider introductory algebra concepts using *Algebra Lab Gear*™. The conceptual base for the model of adding and subtracting integers using this material (area model) is a variation of the colored chip model considered previously.

The anticipated outcome for this component of the course is to build a deeper understanding of what constitutes number sense. This includes the ability to—

• Compose and decompose numbers; move flexibly among different representations; recognize when one representation is more useful than another.

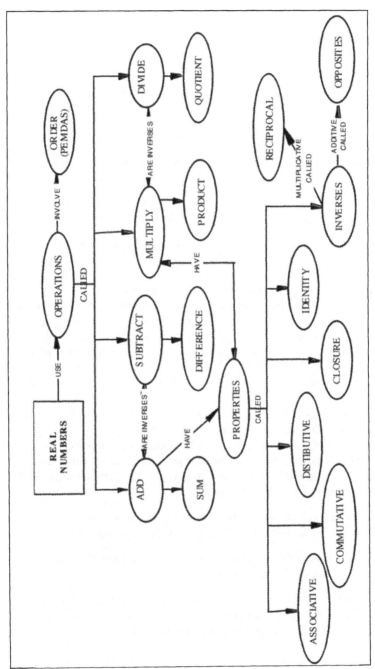

Figure 3: Partial Concept Map about Real Numbers

- Recognize the relative magnitude of numbers.

- Deal with the absolute magnitude of numbers.

- Use benchmarks.

- Link numeration, operation, and relation symbols in meaningful ways.

- Understand the effects of operations on numbers.

- Perform mental computation through "invented" strategies that take advantage of numerical and operational properties.

- Use numbers flexibly to estimate numerical answers to computations and recognize when an estimate is appropriate.

- Demonstrate a disposition toward making sense of numbers (adapted from Sowder, 1992, pp. 4-5).

The core concepts of algebra (Fey, 1990) are considered—variables, functions, relations, equations, inequalities, and rates of change. Explorations are concentrated on examining relations among several quantities whose values change. Students are given opportunities to use graphing calculators and other algebraic software, such as *Graphing Equations and Green Globs*™, with emphasis placed on the use of the multiple representations of tables, graphs, verbal rules, and equations.

Algebra, as a set of formal procedures used for transforming expressions and solving equations, receives secondary emphasis and a problem-centered focus. Students are encouraged to connect their work with algebra with concepts developed earlier when working with real numbers. As with number sense, attention is given to defining what constitutes symbol sense; this includes the ability to—

- Scan an algebraic expression to make rough estimates of the patterns that would emerge in a numeric or graphic representation.

- Make informed comparisons of orders of magnitude for functions with rules of the form $n, n^2, n^3 \dots$ and n^k.

- Scan a table of function values or a graph or interpret verbally stated conditions and identify the likely form of an algebraic rule that expresses the appropriate pattern.

- Inspect algebraic operations and predict the form of the result, or, as in arithmetic estimation, inspect the result and judge the likelihood that it has been performed correctly.

- Determine which of several equivalent forms might be most appropriate for answering particular questions (adapted from Fey, 1990, pp. 80-81).

Issues and Challenges

UNC's Middle Grades program has been in place for some time and is recognized as being innovative with respect to addressing the special needs of middle grades teacher education. For students concentrating in mathematics, both the methods component and the student teaching component are in place. However, as we consider the need to help students integrate knowledge of three central components to teaching (mathematics content, mathematics pedagogy, and students' thinking about mathematics), we have decided that some *tinkering* with an already successful program is necessary (Figure 4).

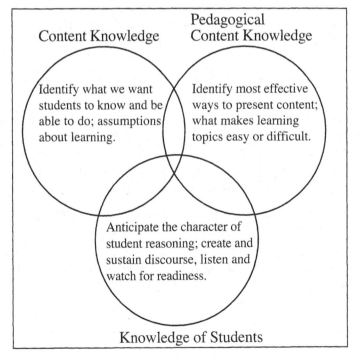

Figure 4: Three Components of Teacher Knowledge

Perhaps part of the dilemma and confusion about teacher education in mathematics is the failure to distinguish between the three components noted in Figure 4. It is not reasonable to address all three ideas at once (as often is attempted in methods course assignments). The development of two new content courses that focus on the mathematics of middle school reflects our efforts to link the mathematics knowledge of students to the content of the middle school curriculum. With these experiences in place, a course that addresses pedagogy and children's thinking with a major focus on the selection and design of appropriate mathematical tasks is a reasonable next step. Finally, during student teaching, attention is given to implementation of and experimentation with knowledge of content, as well as knowledge of pedagogy in terms of planning for teaching and exploring middle grades students' thinking. Clearly, these learning opportunities need to continue for the students when they become practicing teachers; thus, the preservice program is intended to lay the groundwork for continuing development and growth in teaching mathematics.

References

Bartels, B. H. (1995). Promoting mathematics connections with concept maps. *Mathematics Teaching in the Middle School, 1*(7), 542-549.

Bassarear, T. (1997). *Mathematics for elementary school teachers.* Boston, MA: Houghton Mifflin.

Fey, J. T. (1990). Quantity. In L. A. Steen (Ed.), *On the shoulders of giants: New approaches to numeracy* (pp. 61-94). Washington, DC: National Academy Press.

Lappan, G., Fey, J. T., Friel, S. N., Fitzgerald, W. M., & Phillips, E. D. (1998). *Prime time.* Palo Alto, CA: Dale Seymour Publications.

National Council of Teachers of Mathematics. (1991). *Professional standards for teaching mathematics.* Reston, VA: Author.

Sowder, J. T. (1992). Making sense of numbers in school mathematics. In G. Leinhardt, R. Putnam, & R. A. Hattrup (Eds.), *Analysis of arithmetic for mathematics teaching* (pp. 1-51). Hillsdale, NJ: Lawrence Erlbaum Associates.

University of South Florida

Denisse R. Thompson and Michaele F. Chappell
University of South Florida

The Institution

The University of South Florida is a large, comprehensive, public university with four campuses and an enrollment of over 36,000 students. It is the thirteenth largest university in the country and houses the largest College of Education in the southeastern United States.

The Program

The Mathematics Education program is a joint program between the College of Education (which teaches the content-specific pedagogy courses) and the College of Arts and Sciences (which teaches the content courses). At present, there is no separate middle grades program for mathematics education; rather, all prospective secondary mathematics teachers complete a program designed to provide certification for grades 6–12. Consequently, preservice teachers participate in courses that focus on middle grades issues as well as on those that focus on high school issues. Because the program covers 6–12, all prospective middle school and high school teachers receive the same mathematics background, differing only slightly from that of a mathematics major.

The current program contains the following content and content-specific pedagogy courses (with the credit in semester hours indicated). The mathematics methods courses are taught with a strong emphasis on an active, hands-on approach to learning mathematics.

Mathematics

Calculus I (4)	*Elementary Abstract Algebra* (3)
Calculus II (4)	*Geometry* (4)
Calculus III (4)	*Elementary Number Theory* (3)
Discrete Mathematics (3)	*Computer Applications in Mathematics* (3)
Introductory Statistics I (3)	*Early History of Mathematics* (3)
Linear Algebra (3)	

Mathematics Education

Reading the Language of Mathematics (2)	*Senior Internship* (10)
Computers in Mathematics Education (3)	*Senior Seminar* (2)
Teaching Mathematics in the Middle Grades (3; 15 hours field-based)	
Teaching Senior High Mathematics (3; 20 hours field-based)	

The Accomplishments

As a result of participation in the MIDDLE MATH Project, we made changes in the mathematics education courses to reflect issues raised and discussed at the conferences. In the remainder of this case study, we provide details about some of those changes.

Reading the Language of Mathematics examines communication in the mathematics classroom (including reading, writing, and oral communication) as well as general issues about curriculum and assessment. After participation in the first MIDDLE MATH conference, we incorporated a curriculum evaluation assignment into this course (see Appendix I for the assignment guidelines). The purpose of this assignment is to help students learn to evaluate curriculum materials with a critical eye toward ensuring that necessary content and pedagogical style are effectively integrated. A cursory look at published curriculum materials suggests a wide range of interpretation about the meaning and intent of the *Curriculum and Evaluation Standards for School Mathematics* (National Council of Teachers of Mathematics, 1989). Hence, it seems prudent that beginning teachers learn to examine curriculum materials to determine their mathematical substance and potential classroom use.

Another change in this course involved formalizing an assignment related to the incorporation of literature into mathematics classrooms. Each student reads a different book about mathematics and then shares a summary of the book and its possible use in the classroom with other preservice teachers (see Appendix II).

Changes in course assignments and materials were also undertaken in the course *Teaching Mathematics in the Middle Grades*. In the past, we had used a research-oriented text geared to curricular and pedagogical issues in the middle grades. However, it seemed that many preservice teachers lacked the broad frame of practical experience needed to benefit from such a text. Hence, we switched texts and began using the *Curriculum and Evaluation Standards* and the *Middle Grades Addenda*

Series as the main texts for the course (rather than as supplements). Students responded well to the *Addenda Series* because the individual books contain a wealth of activities, tasks, and investigations.

As a complement to these *Addenda* resources, we have students complete an activity file containing resources in each of the major areas identified in the *Standards* (see Appendix III). The activity file has two main purposes. First, it enables preservice teachers to develop a wide selection of resources useful as they begin teaching—resources that are of personal interest to them, and therefore, more likely to be used. Second, developing the activity file acquaints preservice teachers with information about where to find innovative ideas after they enter the middle school classroom.

The other major change in this course was the introduction of a performance exam in the form of an oral, one-on-one interview with the instructor. Preservice teachers are expected to demonstrate their proficiency using manipulatives to explain mathematics concepts and students are provided with a list of tasks to be assessed (see Appendix IV). Different groups of tasks are placed on color coded cards and students must randomly choose cards of each color, then demonstrate proficiency with each set of manipulatives.

This assessment has the benefit of allowing the instructor to address misconceptions if and when they occur. Further, it requires students to consider the language and actions needed to explain mathematics concepts at a developmental level. In retrospect, this form of assessment is no more time consuming than grading a typical exam.

Issues and Challenges

Changes in the assignments for these two courses resulted in an increased understanding of our students' knowledge about the mathematics they will be expected to teach and about their ability to make instructional decisions required of teachers. We hope to continue to use and modify these assessments because they complement written exams and give us a broader means by which to evaluate our students. However, at times, the full benefit of these assignments is not realized because of the limited early field experiences of students— incorporating a broader range and diversity in field experiences remains an on-going challenge.

Another issue and challenge we face is the integration of technology into the program. Although there is a course on computers in mathematics education and we demonstrate graphing calculator technology in the methods courses, not enough time is devoted to developing competence with these tools. We are presently in the process of adding a course to the program that will focus on the use of graphing calculators, calculator-based labs, and symbolic manipulators. We want students to develop competence in their ability to use these tools and consider a wide range of pedagogical issues which arise with their use.

Finally, we need to consider whether the 6–12 program is best suited for all of our students. Occasionally, there are students who struggle with some of the mathematics that is required in the 6–12 program, but that same mathematics might not be required in a 5–8 or 5–9 program. Although a state-level middle grades mathematics certification exists, we have no program leading to certification for only the middle school level.

Appendix I: Curriculum Assignment

In this assignment, I want you to evaluate a textbook with a critical eye, making the types of judgments that would be required if you were responsible for deciding which book to use. You will be trying to gauge the extent to which the textbook seems to be aligned with the NCTM *Standards* as well as with the issues we have discussed in class. In addition, this is an opportunity for you to see the extent to which textbooks have changed since you were in secondary school.

1. Choose a book from the secondary (6–12) curriculum, below the level of calculus, with a copyright date since 1990. Make sure you choose a book that was designed for a secondary audience rather than a college-level text used at the secondary level. You need to obtain a teacher's edition of the book you are considering.

2. Choose one chapter (not the first or last) on which to complete a detailed analysis. A chapter midway through the book should contain new content rather than just review material; reviewing such a chapter should help you get a sense of how the authors introduce new material. Look carefully at the mathematical content, the organization and structure, the types of student experiences, the nature of the teacher's role, and the types of assessment to be used. For this chapter, rate *each* of these areas on a scale from 1 (low) to 5 (high). Provide a rationale, with exemplars when necessary, to support your ratings; your rationales might be on the order of a paragraph each.

Use the *Guidelines for Selecting Instructional Materials in Mathe-matics Education* (National Council of Supervisors of Mathematics, 1993) to provide meaning to each of these categories.

3. After you have looked at one chapter in detail, skim the remainder of the book, looking at the same categories as in #2 above. It will be helpful to have studied the individual chapter because you will be more familiar with features to look at briefly throughout the re-mainder of the book. Again, rate the book as a whole in each of these categories on a scale from 1 (low) to 5 (high). Provide a ra-tionale for your ratings. Your ratings for an individual chapter and for the book as a whole may be the same or quite different, depend-ing on the extent to which you feel that the sample chapter is repre-sentative of the book in its entirety.

4. Provide an overall recommendation concerning the book. If you were on the adoption committee, would you recommend that it be strongly considered for adoption? Should it definitely not be con-sidered for adoption? Justify your recommendation with the high-lights that you would share or point out to other committee mem-bers.

5. Be sure to include the appropriate bibliographic citation—title, authors, publisher, and copyright date. Your paper should be typed.

Grading: Chapter ratings and rationales   30
          Overall book ratings and rationales 15
          Recommendation and rationale   5
          TOTAL   50

Appendix II: Reading Assignment

I.   Choose a book from the *Connecting Literature and Mathematics: Secondary Bibliography* reading list. You may choose to read a book not on the list with prior approval.

II. Sign up for this book on the appropriate sign-up sheet. Everyone should read a different book. The sign-up guarantees that there are no duplicates. We use a FIRST-COME, FIRST-SERVED policy.

III. Prepare a 3–5 page double-spaced paper that contains the following sections—

   a) Complete bibliographic citation (title, authors, copyright, pub-lisher).

b) A brief summary of the book that highlights the essential fea-
    tures—at most 2 pages.

c) An indication of the appropriate audience—middle school, high
    school, teacher (as in teacher resource or reference book). You
    should provide a brief rationale for your decision on audience,
    perhaps citing strengths or weaknesses of the book for the given
    level.

d) An indication of how this book might be used with a class—at
    most 1 page.

e) Your personal reaction—what did you think of the book—
    interesting, useful, etc. Provide some rationale for your com-
    ments. In other words, why did you like or dislike the book?

f) A 1 page single-spaced abstract containing item (a), an abbrevi-
    ated version of item (b), the intended audience from item (c),
    and an abbreviated version of item (d). You should make enough
    copies of this abstract to share with the other members of the
    class.

IV. Your paper should be typewritten. You should use appropriate
    grammar, proper spelling, and proper punctuation. Use a spell
    checker or have someone proof read your paper.

Grading:	Bibliographic citation	5
	Audience and rationale	5
	Personal reaction and rationale	5
	Summary	20
	Classroom use	10
	Abstract	5
	TOTAL	50

[For a copy of the bibliography contact Denisse Thompson or
Michaele Chappell.]

Appendix III: Activity File

One of the difficulties in being a first-year teacher is that you have very
few resources. The purpose of this assignment is to alleviate some of
that problem by having you develop a resource file with activities
relevant to the content of this course and the NCTM *Curriculum
Standards* for the middle grades.

Your task is to collect good, creative activities and resources that would be appropriate for use with middle school students. The activities might be articles that you find in a journal or unusual activities that you find in a teacher resource. THEY SHOULD NOT BE SKILL-DRIVEN WORKSHEETS. Activities that I give you in class or activities from the *Addenda Books* should NOT be included in the activity file that you submit. Also, do not get all your resources from the same source. Use this opportunity to find out where to obtain innovative ideas. Once you are a practicing teacher, you want to know where to look in a hurry to find what you need. Be sure to abide by copyright regulations.

You may feel free to share ideas with each other. Teachers need to learn to share ideas that work.

You may organize the file in a manner that is useful to you. However, I want to see evidence that you have collected activities in the following areas:

- Arithmetic operations—whole numbers, rational numbers (fractions and decimals), estimation
- Problem solving
- Pre-algebra/algebra
- Geometry
- Measurement
- Probability/statistics
- Miscellaneous—multicultural connections, literature connections

Grading:

A (45–50 points). At least 40 good activities balanced among the categories, at least five different resources, and evidence of good organization.

B (40–44 points). At least 35 activities balanced among the categories, at least four different resources, and evidence of good organization.

C (35–39 points). At least 30 different activities balanced among the categories, at least three different resources, and adequate organization.

How valuable a resource this is will depend on the energy and effort that you put into it. If your grade is not at least a C, I will expect you to redo your file. Just having a set number of activities is not enough for a given grade. Your file is also evaluated on the quality of the activities, the nature of the resources, and its organization.

Your file should include a Table of Contents for each category and a list of the resources that you used. Possible resources include *Teaching Children Mathematics*, *The Mathematics Teacher*, *Mathematics Teaching in the Middle School*, *The Arithmetic Teacher*, yearbooks from NCTM, teacher resource materials, and Internet web sites.

Appendix IV: Performance Exam

You should be prepared to demonstrate competence on the following tasks with the indicated manipulatives. During your individual interview you will select several activities at random from this list. I will conduct the selection in such a way that it is not possible to avoid any category. Plan on 15-20 minutes for your interview.

Pattern Blocks

1. Illustrate the meaning of a fraction
2. Illustrate how to develop equivalent fraction ideas
3. Illustrate how to add fractions, with or without a common denominator
4. Illustrate how to subtract fractions, with or without a common denominator
5. Illustrate the meaning of division of fractions

Decimal Grids

1. Illustrate the meaning of a decimal
2. Illustrate how to develop understanding of equivalent decimal representations
3. Illustrate ordering of decimals
4. Illustrate division of decimals
5. Illustrate multiplication of decimals

Tangrams

1. Illustrate relationships between pieces
2. Illustrate spatial activities that can be completed with the pieces
3. Illustrate number relationships based on values of pieces and their interrelationships
4. Illustrate a proof of the Pythagorean Theorem

Geoboards

1. Illustrate geometric figures
2. Demonstrate the meaning of area
3. Demonstrate the meaning of perimeter

Algebra Lab Gear and/or Two-Color Chips

1. Illustrate addition of integers
2. Illustrate subtraction of integers
3. Illustrate simplifying expressions
4. Illustrate multiplying of polynomials
5. Illustrate division of polynomials
6. Illustrate solving of equations

Miscellaneous

1. Use blocks to illustrate the meaning of volume
2. Use blocks to illustrate the meaning of surface area
3. Use blocks to develop formulas for the volume of a rectangular solid
4. Use paper to demonstrate the theorem about the sum of the measures of the angles of a triangle
5. Use paper to demonstrate the expansion of $(a+b)^2$ or $(a-b)^2$
6. Use straws or spaghetti to illustrate the Triangle Inequality Theorem
7. Use a MIRA to illustrate finding lines of symmetry
8. Use play money to illustrate division of whole numbers

Each task will be worth 10 points. If you select a task that you are unable to complete, you may choose another task but with a three-point penalty assigned to the new task.

University of Wisconsin-Oshkosh

Asuman Oktac and Jennifer Szydlik
University of Wisconsin-Oshkosh

The Institution

In its 125 years, the University of Wisconsin–Oshkosh has evolved
from a teacher-training institution to an exemplary, four-year, comprehensive university. Located in the Fox River Valley of central Wisconsin, it is the state's foremost teacher-training institution, producing
thousands of teachers and contributing to Wisconsin's leadership in
education. Serving approximately 11,000 students, the University's four
colleges and graduate school offer a range of programs in education
and human services, letters and science, business administration, and
nursing. However, elementary education remains the most popular
major at the university.

Although there are no faculty with doctorates in math education in the
Mathematics Department, there are five faculty who are very interested
in the area, including Asuman Oktac, who earned a master's degree
with an emphasis in math education, and myself, whose dissertation
involved a research study in math education. There are, however, two
mathematics educators in the College of Education. Generally, in the
program for the preparation of mathematics teachers, mathematicians
teach the content courses and math educators teach the methods courses
and supervise the student teachers.

The Program

At Oshkosh, the middle school program is part of the K–8 certification.
The Wisconsin Department of Public Instruction mandates 12 credits of
mathematics and mathematics methods for every prospective elementary teacher. In addition, elementary education students completing a
degree in the School of Education are required to choose an emphasis
in a content area. Currently, over 80 students have a mathematics
emphasis. These students will be hired either as mathematics specialists
in an elementary school or as middle school math teachers.

Since the Mathematics Department developed the math emphasis
program in the early 1970s, it has been in need of revision. In the

course of the 1995-96 academic year, a new program was developed and is working its way through the university committees for approval. This case study is a description of the new program and the story of its revision process—a story of vision, cooperation, confusion, procrastination, and resolution.

We require of all elementary education students three mathematics courses: *Number Systems*, *Geometry and Measurement*, and *Data Exploration and Analysis*. In addition, in the newly developed program, students with a mathematics emphasis are required to take a three-credit-hour Senior Seminar and three electives. They may choose from *Probability and Statistics*, *Infinite Processes*, *Modern Algebra*, and *Modern Geometry*. These are all four-credit courses and have *College Algebra* as a prerequisite. All of the courses in the emphasis are exclusively for elementary education students and thus allow for discussion of teaching related issues. The students also take *Teaching Pre K–8 Mathematics* and *Student Teaching in the Middle/Junior High School* in the School of Education.

The Accomplishments

We attended the MIDDLE MATH Conference in August of 1995 and examined demonstration units for five new NSF-funded curriculum projects for middle school students. These projects demand of the teacher a broad content knowledge, an understanding of the structure and practice of mathematics, hands-on pedagogy, and flexibility with both calculator and computer technology. We were excited by the projects and committed to creating a program at Oshkosh that would produce knowledgeable, creative teachers who are prepared to implement innovative projects in their classrooms.

In fall 1995, we gave a colloquium at Oshkosh describing the new curriculum projects and invited anyone who was interested to join us in the revision of our program. Both mathematicians and mathematics educators joined our committee, and we met to discuss changes. Seeking the cooperation and help of all interested parties was vital to the success of our effort—neither the Math Department nor the Department of Curriculum and Instruction faculty were surprised by the final proposal or felt that they were not informed or consulted.

Our committee of twelve members articulated the following vision for the ideal outcomes of our program. Our students will—

- Understand the central ideas of number, number systems, number theory, geometry, probability, statistics, discrete mathematics, algebra, and intuitive calculus.

- Communicate mathematics effectively, both orally and in writing.

- Be versatile and confident problem solvers.

- Use both technology and manipulatives to enhance their own knowledge of mathematics and that of their students.

- Know and appreciate mathematics as an art.

- Have a historical perspective of the mathematics they study.

We reached consensus on most of the vision immediately. Consensus on the method to achieve the vision was not so easily attained. We were constrained to a total of 24 credit hours and 9 of those were already assigned to the three courses required of all elementary education students. Several questions emerged from our discussion:

- Should the remaining 15 credits be devoted to courses on technology, problem solving, or history of mathematics in addition to content courses, or should the courses all emphasize content and incorporate these features as a means for learning that content? If we chose the latter strategy, how could we be assured that technology, problem solving, and history stayed a part of the courses over time?

- Should there be a capstone course for the minor? What would be the focus of such a course?

- Should ideas from the calculus be a part of our offerings or should we focus on discrete mathematics?

- Should we continue to teach courses for this population separate from the courses in the major?

- What technology should be required?

- Should we require a prerequisite for the emphasis program? If so, which course(s)?

There were differences of opinion on most of these issues, and often it was easier to procrastinate than work on the program. Nine months passed.

In the end, the commitment to revision triumphed over differences in philosophy among committee members and, in fact, our numerous discussions of the issues served to clarify positions and to provide organization out of confusion. We reached a compromise on all the previously mentioned questions, with the exception of the inclusion of calculus. In that case, the committee voted and the majority opinion (to include calculus ideas) was honored.

We chose to offer four elective courses: *Probability and Statistics*, *Infinite Processes*, *Modern Algebra*, and *Modern Geometry* from which students will choose three. These four credit courses integrate problem solving, cooperative learning, technology, and history with their content. Inclusion of these features is supported by requiring a history text and a graphing and statistics calculator (the TI-83) of all students in the emphasis program. We have incorporated several *Minitab* activities into our *Probability and Statistics* course and plan to create computer assignments for the other courses. Additionally, the instructors for the courses are committed to developing packets of interesting problems to serve as group work. We received a grant to collect and write the problem packet for *Probability and Statistics*, and we completed that work in the summer of 1996. *Infinite Processes* uses a packet (*Concepts of Calculus for Middle School Teachers*) prepared by Maier at the Curriculum Development Lab of Portland State University, and *Modern Algebra* requires a packet (*Algebraic Structures*) prepared by Koker, an Oshkosh faculty member. We are currently collecting materials for *Modern Geometry*.

In addition to 12 credits of electives, students are required to complete a three-credit capstone course entitled *Senior Seminar*. The class includes a survey and study of the research literature focusing on the teaching and learning of upper elementary and middle school mathematics, emphasizes connections between the courses in the emphasis area, and provides hands-on experiences with elementary and middle school curriculum materials. Students prepare activities from the materials and model the use of their activities in the class. Relevant mathematics is discussed.

Our courses strike a balance between hands-on experimentation (inductive thinking) and structured arguments (deductive thinking). For example, in *Modern Algebra* we study the structure of algebraic systems inductively by flipping and rotating triangles, adding on clock faces and number lines, and creating operation tables for mystery sets and operations. We also study the structure of algebraic systems deductively by writing proofs based on the definitions of a group and a ring. In *Probability and Statistics*, students perform experiments by flipping coins, rolling dice, and playing mathematical games. They collect, display, and analyze data and report their findings. They also provide arguments for important results (e.g., the number of committees of size k that can be formed from a group of n members is $n!/((n-k)!k!)$). The content is flexible enough to include topics of the students' or instructor's choosing. For example, in *Modern Geometry* topics such as fractal geometry and topology may be included.

Issues and Challenges

Our program requires growth from us as instructors. Our pedagogical emphasis is on problem solving and class discussion of the problems. We believe that students learn mathematics best by doing mathematics each day in the classroom. Our program places responsibility on the instructor to compile or create interesting, content-rich problems, to lead class discussion, and to respond effectively to student arguments. This is an ongoing challenge. It requires us to learn to use technology as a teaching tool, and to develop or adopt activities that support student use of that technology. Although we know how to employ technology to teach ideas in calculus, we need to learn how to use it to teach geometry, statistics, probability, and modern algebra.

Western Carolina University

Kathy Ivey and Scott Sportsman
Western Carolina University

The Institution

Western Carolina University is a public, comprehensive, four-year university located in the scenic mountains of Western North Carolina. Founded in 1889, WCU is the westernmost of the 16 senior institutions of the University of North Carolina system. Total enrollment is approximately 6,700 students, but WCU serves about 12,000 people annually through credit and non-credit courses, continuing education offerings, workshops, and conferences.

There are mathematics education faculty in both the Mathematics Department and the College of Education and Allied Professions. In the program for the preparation of teachers of middle grades mathematics, some of the mathematics content courses are taught by mathematics education faculty. Student teachers at the middle grades level are not necessarily supervised by mathematics education faculty — and definitely not by faculty from the Mathematics Department.

The Program

The middle grades program is separate from, but closely connected to, the elementary program. It provides in-depth study of topics from elementary and middle grades mathematics and a broader perspective on the mathematics of high school and early college. Courses in the mathematics focus area include the *Theory of Arithmetic I* and *II*, *Trigonometry*, *Algebra and Analytic Geometry*, *Introductory Calculus* or *Calculus I*, *Applied Statistics*, and *Informal Geometry*. Students must also choose one course from *Calculus II*, *Statistical Methods I*, *BASIC Programming* or *Computer Science I*.

The Accomplishments

Although we have not made changes in the actual program for preparing teachers of middle grades mathematics, we have changed the type of mathematics preparation these preservice teachers receive. The NCTM *Standards* (1991, 1989) present a vision of public school mathematics in which students are actively involved in explorations. Based on our belief that teachers, particularly new teachers, tend to teach in ways they have been taught, we decided to try to include more active explorations in prospective teachers' mathematics classes. We already include mathematics lab activities in our calculus sequence, but our middle grades mathematics education students may not take that sequence. Therefore, we decided to add some math lab activities to our pre-calculus course. This case study will describe our experience over the last year of using math labs in pre-calculus. We begin with our rationale for choosing pre-calculus and with a general description of our math labs. Then, we discuss some advantages and disadvantages of these activities and our plans for the future.

Although the focus of this report is on prospective teachers, our desire to incorporate labs into mathematics classes was more general. We hoped that all students would benefit from an active approach to mathematics and from writing lab reports. Additionally, we wanted our prospective teachers to see a different approach to teaching mathematics than the traditional direct instruction presented in a lecture. We chose pre-calculus for three reasons. First, we both requested sections of pre-calculus to teach and were willing to work together to develop lab activities for these classes. Second, pre-calculus is the entry-level course for many of our students and required for all students with an elementary or middle grades mathematics concentration. Third, we hoped that adding lab activities to the pre-calculus course would help our students prepare for our calculus sequence. We currently use lab activities in the calculus sequence to supplement the Harvard Consortium reform textbook (Hughes-Hallett et al., 1994).

All our math labs have a common goal: creating experiences for our students to share which we can then use as basic examples throughout the discussion of each major topic or that we can use to tie together several topics. Consequently, we tried to create labs that could be done at or near the beginning of a major topic of discussion or that served as a comprehensive review. Using these labs as a basis for examples in later class discussions is a particularly important point. Some of our labs involve data collection and analysis using a Calculator Based Laboratory from Texas Instruments (CBL) and a graphing calculator. We have used CBL activities from several sources as starting points for developing labs that fit our particular needs more closely. (See for example, Brueningsen, Bower, Antinone, & Brueningsen, 1995; Nichols, 1995.) All of our students are required to have a TI-85 graphing calculator for our pre-calculus and calculus sequences, but many of our students begin with little or no knowledge of the operation and limitations of a graphing calculator. Therefore, we also try to train students to use their calculators effectively as tools to explore mathematics. Since 1997, we have required the TI-83 because it is more user friendly.

Currently, our pre-calculus course covers the following major topics: functions with special emphasis on polynomials, exponentials, and logarithms; conic sections; and sequences and series. Students complete five labs during the pre-calculus course. The first lab begins on the first day of class. We have students take a variety of measurements, including height, arm span, hand span, arm length, head circumference, wrist circumference and other similar measurements. This lab has three purposes. First, it sets the tone of active class participation using student-collected data. Second, it allows students to learn the names of several classmates and to begin to feel like an integral part of a group. Third, it produces data that we will use over the next few days to review and/or to introduce linear equations. Using these data sets, students produce eyeball linear fits that allow us to discuss the computation of slope, how slope affects the "tilt" of the line, how to shift lines vertically, and finally, how to write and read linear equations. We also discuss the idea of goodness-of-fit and introduce residuals for assessing the accuracy of predictions, based on their equations, for points not in the data set.

After an extended discussion of functions in general, we turn to polynomials beginning with quadratics. Our second lab uses the CBL and a Vernier Ultrasonic Motion Detector. We have students, in groups of four or five, toss a beach ball straight up over a detector placed on the floor and then allow the ball to fall back down. They also perform the same procedure with a balloon. Finally, they hold a beach ball high over the detector and drop it. For the first two activities, the students fit a quadratic equation to the data by using the vertex and another point to determine the coefficients. We have them choose several different points and average each set of coefficient values for a final fit. For the balloon, they can usually see the drag associated with wind resistance in their graphs. They predict when the ball and the balloon hit the ground (the detector will not record data within two feet of it) and use residuals to assess the accuracy of their predictions. For the dropped ball, students choose three points and solve a system of equations to determine the coefficients of the quadratic fit. Again they use several sets of points and average their final values for the coefficients. The lab has three basic purposes. First, we use this lab to discuss roots of a quadratic and how those roots are related to the standard form of the equation. Second, the lab gives students practice in working with different forms of the quadratic equation, and third, it leads to a more general discussion of translations of graphs. We generalize the concepts developed for the quadratic to examine roots of polynomials in general and translations of graphs of polynomials.

Our next major topic is exponential and logarithmic functions. We begin this part of the course with lab three. We have two main purposes for this lab. First, we want students to see an example of exponential-like data in a concrete form. Second, we want students to experiment with varied bases for the exponential and to learn the effects of the various constants in the standard exponential equation $y = A(B^x) + D$. In this lab, we have students drop large and small diameter balls under-neath the Ultrasonic Motion Detector and collect data about the succes-sive return heights for five or six bounces. Using only the high point of each bounce, we have students fit a general exponential equation to their data. We begin with a small ball, such as a racquetball or a golf ball, so that the diameter of the ball is negligible. In this way, we can assume a horizontal asymptote of $y = 0$ with little effect on the fit of the data. Students determine an appropriate coefficient and then, through guess and check, determine an approximate value for the base of the exponential. We also introduce the idea of using the data to calculate the percentage of return as another approximation for the base. Once again, students are asked to predict a value for some point not in the data set and use residuals to assess their accuracy. They repeat this analysis with a large diameter ball, such as a basketball or a child's kick ball. This time, they must account for a horizontal asymptote in their equations. For a different view of the legitimacy of using an exponential model for this activity, see Barbeau (1996).

Our fourth lab is intended to provide connections among all types of functions examined to this point in the pre-calculus course. In this lab, we ask students to walk in front of the Ultrasonic Motion Detector and CBL to generate graphs of different types of functions. One purpose of this lab is for students to gain a kinesthetic knowledge of the behavior of different types of functions—which ones grow or decay rapidly, which ones change slowly, which ones dominate, which ones have asymptotes, and which ones have similar end behavior. Then, students are asked to examine the behavior of families of functions by system-atically varying the constants in the equations. A second purpose for this lab is for students to describe the important features of these families of functions.

Our final lab, the fifth, examines conic sections in a very traditional way with a new twist. Traditionally, most of us either describe how the different conic sections can be seen in the intersection of a plane and a double-napped cone or show a wooden or plastic model of the various cuts. In the fifth lab, we have students cut Styrofoam cones (found in the artificial flower section of a discount store) to create their own model of the various conic sections. Once these sections have been cut, students trace the outlines of the intersections onto graph paper and determine an equation for each one. Our purpose here is to make the various standard measurements in the graphs of conics have concrete analogs. In doing this activity, students work directly with translated equations of conic sections. The location of the vertex and the center, as well as the lengths of major and minor axes and the position of the foci, becomes problematic, particularly for finding the center of the hyperbola. In figuring out the equations, students deal with all the typical lengths associated with conic sections, but with a more concrete use for those numbers.

We have made several changes in our pre-calculus course to accommodate the labs. We changed the meeting time from three fifty-minute classes per week to two seventy-five-minute classes each week. The longer class period allows us to complete a lab in one day. The dean of our college has allowed three classrooms to be reconfigured by removing fixed desks and replacing them with moveable ones. We have changed the order of presentation of some topics and the types of questions that we ask students. A typical exam is mostly context embedded questions that can be solved in several different ways. For example, rather than ask students to find the roots of a particular polynomial, we may ask them to fit a set of data with a given type of equation, use their fit to predict when a root will occur, and assess the accuracy of their prediction. Most of these changes required the support of both the department and the college administrations. We have been fortunate to have this support.

Issues and Challenges

In this section we discuss some advantages and disadvantages of the use of labs to enhance the teaching of mathematics. One major advantage of labs is the increased student interaction. All of our labs are group activities—our students work together in established groups almost every class period. Many of these groups meet outside class to work homework problems and to study for tests. After we started using labs, we noticed that student groups became more interactive during regular classwork.

Students transferred their ownership of the lab problems (their walking or cutting the cone) to problems from the book. They seemed more willing to accept the situations or data given in a book as representative of a "real" situation.

A third benefit of the use of labs is that students engaged in more authentic mathematical communication. There is more discussion between students focused on the concepts and procedures rather than simply the mechanics of a particular calculation.

There are indications that the kinesthetic experience of the labs has a deep impact on some students. Their understanding of mathematical concepts seems to be closely tied to their physical experiences. This is one area about which we are curious and plan to explore in more depth.

Finally, the labs have also increased faculty interaction. We are frequently in each other's classes and our students are willing to ask either of us for assistance. We plan our classes together and create tests together. Because we often approach topics and problems from different viewpoints, we both have a richer basis for our teaching. Other faculty members have used some of our labs in their classes and have asked for assistance in developing labs of their own.

Potential disadvantages include preparation time, initial student reactions, administration support, and cost. The major disadvantage to labs is time—not the time taken from class, but rather the preparation time and the time needed for evaluating student learning. We have spent many hours preparing lab sheets, writing programs to run the CBL, and grading lab reports. Having two or more people involved in this process reduces the time needed, but it is still a time intensive approach to teaching. On the other hand, time has not been an issue in covering the designated material from the course syllabus. We find that the labs give our students a common basis of experience from which we can draw examples for illustrations, thus saving us time during the rest of the course.

Another disadvantage to labs is that not all students enjoy or find the lab activities beneficial. Students particularly object to the writing component of the lab reports. Most of these students see mathematics as something to be memorized and are not happy when we ask them to analyze, conjecture, explore, and then synthesize their results into a written report. However, our experience has been that most of our students do benefit from the lab activities.

A related *potential* disadvantage to labs (that we have not experienced) is a negative reaction from administrators. We have had strong support from both our past and current department chairs and from our dean. Attempting this type of change in an established course without such support could prove very difficult.

We are often asked about the financial cost of these labs. Some equipment was borrowed from our physics and chemistry department and we purchased enough CBLs and various probes to outfit seven groups for less than $2000.00. Most of this money came as an internal grant for instructional improvement.

What effect do these labs have on student learning? The answer is — we don't know ... yet. We are currently working on a research project to examine the use of labs in developing students' understanding of the concepts of pre-calculus and calculus. We have just now reached the point at which our students who have come through our lab-enhanced pre-calculus and calculus sequence are entering upper division mathematics classes. We do not know what effect we will see in the rest of their undergraduate work or in how our prospective teachers will approach their own classrooms. A follow-up study of these students is in the planning stage. We would like to interview some of our students as they continue to work towards their degrees and then as they move into public schools. In the short term, we can say that we see more willingness to engage in mathematical discussions in the classroom, more active participation in class, and more student interaction both in and out of class. Both students and faculty find the labs both challenging and useful.

This brief description of our work with math labs should provide some insight into how our program for middle-grades teachers (and for all students) is changing. As public schools move in the direction of the vision of the NCTM *Standards*, we at the university level must begin to prepare our prospective teachers to work and teach in active environments. We hope that the labs we are developing at Western Carolina University are a first step in that direction.

References

Barbeau, E. (Ed.) (1996). Fallacies, flaws, and flimflam: FFF#111. The bouncing ball. *The College Mathematics Journal, 27*(5), 372-373.

Brueningsen, C., Bower, B., Antinone, L., & Brueningsen, E. (1995). *Real-world math with the CBL system.* Texas Instruments.

Hughes-Hallett, D., & Gleason, A. M. (1994). *Calculus.* New York: John Wiley & Sons, Inc.

National Council of Teachers of Mathematics. (1989). *Curriculum and evaluation standards for school mathematics.* Reston, VA: Author.

National Council of Teachers of Mathematics. (1991). *Professional standards for teaching mathematics.* Reston, VA: Author.

Nichols, S. D. (1995). *Explorations in pre-calculus for the TI-82.* Erie, PA: Meridian Creative Group.

Western Michigan University

Christine Browning and Dwayne Channell
Western Michigan University

The Institution

Western Michigan University (WMU) is one of Michigan's five gradu-
ate-intensive, research-oriented, public universities. It has earned
national and international recognition for its teaching, graduate education and research, becoming the only public university in Michigan to
be designated as Doctoral I by the Carnegie Foundation for the Ad-
vancement of Teaching. Located in Kalamazoo, it has an undergraduate
student population of over 20,000. As of 1997, it is the home of four
federally-funded research efforts, totaling over $12 million, to improve
the nation's schools and reshape the way mathematics is taught.

The State of Michigan does not currently offer middle school certification. Instead, prospective middle school teachers must obtain elementary certification with at least a minor in mathematics or secondary
school mathematics certification. The eight mathematics educators in
the university are on the faculty of the Department of Mathematics and
Statistics in the College of Arts and Sciences. There are no mathematics educators in the College of Education. All elementary and secondary mathematics methods courses, as well as all mathematics content
courses for the elementary/middle school teaching program, are taught
by the mathematics education faculty or by local pre-college mathematics teachers and graduate students. Content courses in the secondary
teaching program are taught by mathematicians. Our prospective
mathematics teachers take general pedagogical and field experience
course work through the College of Education. Prior to the implementation of the project described in this case study, none of the field
experience work, including intern teaching, was supervised by a university mathematics education specialist. Having all of us housed in the
College of Arts and Sciences has been an obstacle to field supervision
by mathematics educators, a problem we hope to remedy.

The Program

At WMU, certification to teach mathematics in the middles grades can
be obtained through successful completion of either the Elemen-

tary/Middle School Mathematics Teaching Minor or the Secondary
School Mathematics Teaching Major (or minor). Mathematics content
and methods courses for both programs are offered through the Department of Mathematics and Statistics. This case study focuses upon a
required course in the secondary programs. The Secondary School
Mathematics Teaching Major requires a minimum of 40 semester hours
of course work. These specialty courses consist of 31-32 hours of
mathematics content and nine hours of methodology (three hours in the
teaching and learning of middle school mathematics, three hours on the
uses of computing technologies for the teaching and learning of
mathematics, and three hours in the teaching and learning of high
school mathematics). Other pedagogical and clinical course work in
support of this program, totaling a minimum of 31 semester hours, is
offered through the College of Education. Students in this program
must also complete course work in a second discipline and additional
course work in general education. A student's overall course of study
must total at least 122 semester hours to graduate.

In the major program, our goal is to offer a mathematics teacher preparation program where students experience effective mathematics
teaching, acquire a strong and varied mathematics background and an
understanding of school mathematics, understand the pre-college
learner, and acquire a knowledge of a variety of pedagogical techniques
appropriate for directing mathematical learning at the secondary school
level. The program is designed around the assumption that students
need many experiences over a long period of time in order to mature
both as a student of mathematics and as a student of the teaching of
mathematics. In addition, students are required to join the National
Council of Teachers of Mathematics.

The Accomplishments

The MIDDLE MATH Project came at a very opportune time for us. We
were already working with our university colleagues and others in the
Teacher Preparation Component of the Michigan Statewide Systemic
Initiative (MSSI) examining ways to strengthen preservice programs
for secondary and middle school mathematics teachers across the state
of Michigan in general, and more specifically, for us at Western Michigan University. Based upon this work, and the additional experiences
provided by our participation in the MIDDLE MATH Project, we
sought and received funding from the Michigan Eisenhower Higher
Education Professional Development Grant Program for the Secondary
Mathematics Teacher Preparation Improvement Project. We designed

the project to address four components of our preservice program with the intent to strengthen the educational experiences provided to our prospective teachers and to bring the program in line with the recommendations of the MSSI teacher preparation group. These components are—

- The implementation of several major changes in, and coordination of, the content and assessment techniques used in three required mathematics pedagogy courses;

- The development of a pre-intern, middle school field experience that provides mathematics teaching experiences in multicultural settings;

- The development of a minority recruitment and retention program; and

- The continued development of program evaluation techniques.

For purposes of this case study, we will focus upon those features of the first two components that relate to the first of three content-specific, pedagogy courses required of all our secondary school mathematics teacher education majors and minors: *The Teaching of Middle School Mathematics*. Previously, this was a two-credit hour course that focused upon the National Council of Teachers of Mathematics (NCTM) *Curriculum and Evaluation Standards for School Mathematics, Grades 5-8*. Essentially, the course provided our students with a sampling of mathematical topics and hands-on, manipulative-based activities (mostly taken from NCTM publications) that were suggestive of the spirit of the *Standards*. Assessments were very traditional, typically consisting of a midterm, a final, and a few reviews of journal articles. In general, students enjoyed this course and rated it very highly.

When the course was first introduced, we believed it would set our secondary mathematics teacher preparation program apart from others by providing a focus on teaching in the middle school classroom. However, as instructors of this course, we saw the need for an increased emphasis on the interrelationships among content knowledge, pedagogical knowledge, and pedagogical content knowledge; the need to expand upon the types of student assessments used in this course; and the need to provide some content-specific, middle school field experience in conjunction with the on-campus classroom activities. The MIDDLE MATH Project provided us with important information and information sources necessary for an appropriate redesign of this course. The funding from the Eisenhower Grant Program has provided and continues to provide us with the time needed to properly plan, coordinate, design, implement, and assess our program changes.

Our newly designed *Teaching Middle School Mathematics* course, now three credit hours, continues to focus on curricular issues and trends related to the teaching and learning of middle school mathematics. However, our emphasis in the course has broadened extensively and now includes attention to the following major areas—

- *Problem Solving, Reasoning, and Communication in Mathematics:* Emphasis is on non-routine problem solving in small, cooperative group settings along with several written and oral presentations.

- *Sense Making in Mathematics:* Emphasis is on learning rational number concepts and algorithms through the use of manipulatives and meaningful arguments that do not rely on memorized procedures.

- *Children's Thinking:* Emphasis is on a social-constructivist philosophy of mathematical knowledge and learning with some discussion of information processing theory, meaning theory, Piagetian ideas, and the Van Hiele model of geometric understanding.

- *Case Studies of Middle School Mathematics Teaching:* Emphasis is on teaching rational number concepts and assessment.

- *Pedagogical Issues:* Emphasis is on assessment, use of technology, writing, collaborative learning, diversity, teacher and student roles in classroom discourse, and creating learning environments that promote sense-making and reasoning.

- *Review of Reform Curriculum:* Emphasis is on the preservice teacher reviewing current middle school reform curricula and, based upon ideas from these curricula, presenting a peer-teaching lesson appropriate for a middle school mathematics classroom.

- *Field Experience in a Middle School Classroom:* Emphasis is on the preservice teacher focusing on how middle school students make sense of and communicate mathematics.

With regard to the last major item, we have been able to work with a large, local, urban school district to design and implement the content-specific field experience component offered in conjunction with the course. This early field experience work provides opportunities in a middle school setting for our prospective teachers to focus on student thinking and understanding of mathematics, to focus on the determination of learning needs of students from underrepresented groups, to work with a diverse group of children in a multicultural setting, and to interact with middle grades children over an extended period of time.

Piloting and Assessing the Course

We ran a pilot of the revised middle school methods course during our spring 1996 term (7 weeks in May and June) with 8 students. Although condensed and offered to a small number of students, this pilot gave us an opportunity to collect materials, try out projects, work in the classrooms, and provide trial runs of assessments. During the first week of the course, the students wrote what could be labeled a "mathematical autobiography." We had them write about significant mathematical moments in their Kindergarten through current mathematical life, their reasons for wanting to be a mathematics teacher, their thoughts on the ideal mathematics student, and their beliefs on what it means to teach and learn middle school mathematics. During the course, each student compiled a journal based on prompts focusing chiefly on their beliefs and attitudes about the teaching of mathematics and upon their reactions to assigned readings and classroom experiences. Their final journal entries had them respond to the following prompt:

> List and describe two elements of this course which were most valuable to you; list and describe two elements of this course that could be improved upon. Then, state specifically how the change might be made; and list and describe two elements of this course that were intellectually most challenging for you.

The following contrasting entries from the initial autobiography and the final journal assignment from two students serve to emphasize the effect of the course experiences on our preservice teachers' beliefs and attitudes.

From Jerry's autobiography:
> *I think that, in a lot of cases, it's a matter of survival, for both the student and the teacher. ...Teaching middle school math involves getting past the little crises and adjusting to the ebb and flow of percolating hormones (which is comparable to piloting a boat with a hole in its hull in a hurricane in the Bermuda Triangle at midnight) and actually getting some information through.*

From Jerry's final journal entry:
> *After sitting through three years of what has essentially been*

lecture-based, "high-level" material, its difficult to think about
guiding students who are still in what is very much a mathe-
matical discovery stage. Just divorcing myself from that type of
thinking has been, and will continue to be, a huge challenge. ...
In order to break the chain of mediocrity, teachers need to push
their students to reach higher and farther by challenging their
minds and providing them with informational building blocks
with which they can construct further knowledge, rather than
simply showing them a finished house and asking them to de-
scribe it again later. Teachers need to create wonder in their
students and show them that math is more than just magical
formulas and correctly calculated answers. That's what a
teacher has to do: facilitate wonder, learning, and inquiry.

The shift in focus on what it means to teach mathematics for this
student is significant. He initially thought of teaching as getting information through to students during a "stormy" period of life; but, after
participating in this course, his thoughts moved to a more dynamic
view of facilitating "wonder, learning, and inquiry."

From Suzanne's autobiography:
[Middle school students] also find it difficult to concentrate on
one topic for a long period of time and crave variety. However,
they also require structure and routine. A teacher of mathemat-
ics in a middle school needs to create lessons that are well
structured and contain a variety of tasks dealing with the con-
cept to be learned, including a way for the students to demon-
strate what they already know and some method for the students
to summarize what they have learned.

From Suzanne's final journal entry:
Since my math experience has been following rules and exam-
ples (rote learning), the activities and problems that used
alternative conceptual models (e.g. measurement vs. distribution
for fraction division) and those that used manipulatives and dia-
grams to formulate generalizations about concepts were most
valuable because they gave me insight on how to present
mathematical ideas.

This student's initial beliefs about teaching mathematics centered on
structure and routine—one needs to have a predictable lesson format
for the middle school students to learn mathematics. The course provided her with alternative methods for focusing on concepts that she

previously would have presented in a traditional "here's how you do it" manner.

Of course, we are only making inferences from the journals of these two students. As we are able to increase our role in the intern teaching of our students, we hope to watch our preservice teachers in action to see if their words actually reflect their belief structures and their teaching practice. Although we were basically pleased with the comments our students provided in their summary journal entries, we believed that perhaps more substantial changes in their beliefs about teaching would be possible in a standard 15-week semester.

Issues and Challenges

There were many other useful evaluative comments contained in the students' writings. All believed the classroom experiences, although brief, were very valuable. They had forgotten what middle school and middle school students were like, they had not worked with children from cultures other than white middle class, and they didn't know what was in the middle school mathematics curriculum. In addition, they all believed the experience of working with a variety of manipulatives was extremely helpful. Having knowledge of and access to some of the reform materials, such as the NSF-funded curricula presented at the MIDDLE MATH Conference, provided more support for the use of manipulatives in developing mathematical ideas. The students saw how manipulatives could be incorporated into "prepared" lessons and know that such curricula exist for them to use in their own classrooms rather than having to search for or create such activities themselves.

However, not all student comments were positive and supporting. Some students felt that the "reading" load was too heavy. Others, who were introduced to the use of rubrics in assessment for the first time, found them difficult to use and (in one case) an impediment to their writing freedom. These students failed to communicate their concerns with rubrics even though class time was given to raise questions over the criteria. It was unfortunate to read such comments at the completion of the course when we had little or no opportunity to discuss such issues further with these students. It indicated to us that more time should be spent on developing a classroom environment where pre-service teachers feel free to question evaluation guidelines and to share their concerns over the evaluative comments on completed work.

Another problem area was reflected in the quality of the field experience provided for our students. Even though we discussed at length,
and shared in writing, our expectations for the field experience with
participating classroom teachers, what happened was not what was
envisioned. Our intent was for WMU students to be actively involved
in listening, observing, and assisting small groups of middle school
students as they learned mathematics. Instead, they either observed a
whole-class lesson from the back of the classroom or they assisted
students in correcting errors on recently-returned exams. Based upon
these experiences and further discussions, the middle school teachers
have encouraged us to develop a workshop for future participating
teachers, focusing on content and pedagogical content knowledge that
will better enable them to model the type of teaching called for by
reform efforts and to permit more involvement by the WMU students
with middle school students.

We plan to continue to act on the evaluative comments provided us by
our students and the classroom teachers—to modify and improve this
course and our program and to provide our students with experiences
that support an alternative view of teaching and learning mathematics.
We are appreciative for the good start the MIDDLE MATH Project has
given us as we work toward implementation of these improvements.

Wright State University

Susann Mathews
Wright State University

The Institution

Wright State University is a metropolitan university meeting the needs of a diverse population of students in the Dayton, Ohio area. It is an open enrollment, public, state university serving nearly 17,000 students. In general, it is a commuter school and a large proportion of the students have nontraditional backgrounds.

Currently, we have three tenure-track mathematics education faculty in the Department of Mathematics and Statistics, which is in the College of Science and Mathematics (CSM) at Wright State. Eight years ago when education reform became a nationally recognized issue, the College of Education and Human Services (CEHS) became involved with Partners Transforming Education, a program in which the local school system, the university, and the community participated in a model designed to coordinate the simultaneous renewal of the education of educators and the pre K–12 school population. Because of his involvement in this program, the mathematics chairman realized the need for more cooperation between CEHS and CSM. He used this program as a mechanism to get the Dean of CEHS and the Dean of CSM engaged in transforming the education of future mathematics teachers. The mathematics chair hired the first mathematics educator in the Department of Mathematics and Statistics in 1989, paving the way for the hiring of two mathematics educators with dual appointments in CEHS and the CSM in 1993 (one with primary appointment in Teacher Education and the other with primary appointment in Mathematics). Additionally, several instructors, many of whom have taught mathematics in public schools, have been hired in the Department of Mathematics who have a strong background in both mathematics and education. However, although our mathematics educators teach mathematics content courses, we do not supervise student teachers.

The Program

Middle grades preparation is currently part of a K–8 Program; however, it will become its own program as CEHS adopts a professional school model and the Ohio Legislature adopts a Middle Childhood License.

In revising and developing the program, we reconsidered the mathematics concentration for elementary and middle grades teachers in light of the Ohio Standards for Teacher Education. We studied the curricula of middle school mathematics concentrations offered by universities noted for their mathematics teacher preparation programs, and we considered ideas learned at national conferences and workshops devoted to the mathematical preparation of middle grades teachers. In this case study, we describe the highlights of our revised program.

Our program must meet the needs of both traditional students and nontraditional students who want to become middle school mathematics teachers. All preservice teachers in the K–8 Certification Program, no matter what their concentration, must take *Quantitative Reasoning*, one of two courses offered in *Intermediate Algebra*, *Fundamental Mathematical Concepts I* and *II*, and *Classroom Applications of Computers*. Additionally, those earning a concentration in mathematics must take *Geometry*, *Algebra and Functions*, *Probability and Statistics*, and *Problem Solving and Mathematical Modeling*. Each of these courses is labeled *"for Elementary and Middle School Teachers."* Additionally, they also must take two of *College Algebra I* and *II*, *Trigonometry*, as well as *Concepts in Calculus*.

The Accomplishments

We have strengthened the existing mathematics concentration for the elementary education majors. Although this is a concentration for preservice K–8 teachers, we anticipate that mainly those students who are interested in teaching mathematics in the middle school will elect the mathematics concentration. We have added both mathematics and education courses to the concentration, and we have revised existing courses.

All students at Wright State University must take a general education mathematics course. Previously, all elementary education majors took *Mathematics in the Modern World* to satisfy that requirement. Effective in the 1996-1997 school year, all preservice elementary and middle school teachers take *Quantitative Reasoning*, rather than *Mathematics and the Modern World*. *Quantitative Reasoning* has been developed to help preservice teachers develop number sense, an ability to estimate, an understanding of how to read and create tables and graphs for data analysis, and a rudimentary understanding of probability.

All students then take a two-quarter sequence of *Fundamental Mathematical Concepts* (a.k.a. *Mathematics for Elementary Teachers*). Each of these courses has a laboratory for problem solving with an emphasis on using manipulatives to make sense of the mathematics. We have added *Algebra and Functions for Elementary and Middle School Teachers* to the concentration and replaced *Problem Solving in School Mathematics* (a 3-hour education course) with *Problem Solving and Mathematical Modeling for Elementary and Middle School Teachers* (a mathematics course). In the process, we have changed from a course with an educational emphasis to a 4-hour course (all credit hours are quarter-hours) with an emphasis on real-world problem solving that will be the capstone of the new mathematics concentration.

We have strengthened *Probability and Statistics for Elementary and Middle School Teachers* from a 3-hour course to a 4-hour course. We also added a technology component, and the course has become student-centered, taking place in the computer lab and providing more depth in the study of probability and statistics. *Geometry for Elementary and Middle School Teachers* has developed into an inquiry-based course in which the students use the computer software, *Geometer's Sketchpad*. Additionally, we are developing an upper-level course, *Concepts of Calculus,* and we have added *Early and Middle Childhood Mathematics* to complement the existing education course, *Elementary School Mathematics: Curriculum and Materials* — previously the only mathematics methodology course that elementary education majors took. The new course will add both breadth and depth, discussing such issues as the use of calculators and computers, problem solving as a basic skill, connections between different mathematical concepts, and materials from new curriculum projects.

We have been changing the methodology with which we teach all of the content courses for preservice middle school mathematics teachers. The courses are taught in a manner consistent with *The Curriculum and Evaluation Standards for School Mathematics* and the *Professional Standards for Teaching Mathematics* (NCTM, 1989, 1991). We believe that one of the best ways of helping students learn sound mathematics pedagogy is teaching in a manner consistent with the *Standards* and then helping them reflect on that teaching and their learning.

We developed an outline for the mathematics portion of a dual major in mathematics and science for prospective middle grades teachers. This major would have been housed in CSM. It was developed in light of the plan of the CEHS to adopt a professional school model and the Ohio Legislature's plan to adopt a "Middle Childhood License." However, CEHS has decided that prospective middle childhood teachers will earn their undergraduate degree in education. Therefore, we have revised our plans. We are adapting our proposed mathematics portion of the dual mathematics/science major to a strong mathematics concentration in CEHS. This curriculum will contain 20 credit hours of mathematics at the 300-level or above: *Geometry for Elementary and Middle School Teachers*, *Probability and Statistics for Elementary and Middle School Teachers*, *Algebra and Functions for Elementary and Middle School Teachers*, *Problem Solving and Mathematical Modeling for Elementary and Middle School Teachers*, and *Concepts of Calculus for Elementary and Middle School Teachers*. This is in addition to the three mathematics foundations courses. There will be 7 hours of methodology courses. Implementation began during Winter Quarter, 1997.

Issues and Challenges

Many of the changes and additions to previous courses are a direct result of Jeri Nichols and my experience with the MIDDLE MATH Project. Although the general education requirement was already evolving into Quantitative Reasoning, our experiences with the project validated our belief that preservice teachers need a well-developed number sense, an ability to estimate, and an understanding of data analysis and probability. Developing these abilities has become the focus of this course. Seeing examples of the new middle grades mathematics curricula and discussing probability and statistics with participants at the MIDDLE MATH Conferences gave me ideas for what to add to our old statistics course and made me aware of the need to make technology a strong and integral component of that course. Our experiences helped us decide what to include in the new methods course to meet the specific pedagogical needs of future middle school mathematics teachers. These needs go beyond what could be covered in the one mathematics methodology course the preservice teachers previously took. Exposure to new curricula highlighted the need for a strong problem-solving course with an emphasis on mathematical modeling. I discussed this issue with presenters at the first MIDDLE

MATH Conference who shared their course syllabi and experiences with me. Presentations at both conferences convinced me of the need for a *Concepts in Calculus* course. As with the problem-solving course, I plan to be in touch with professors I met through the project who have already taught such a course. Furthermore, all these experiences provided me with the credibility I needed to convince both my mathematics colleagues and mathematics chair of the need for strengthening existing courses and developing new ones.

It may be significant for others to know that the support and leadership of the chair of the Department of Mathematics and Statistics, Dr. Edgar Rutter, has been of great value throughout the evolution of our program for the preparation of teachers of middle grades mathematics. He has believed in the importance of this program and has been building its personnel resources for more than seven years. He has also helped us achieve acceptance of our new courses and program by knowing how to work within the system at Wright State, understanding who needed to know what when and knowing when course proposals needed to go to various committees to get their approval. Finally, he has been invaluable in his understanding that, to form a strong alliance between CEHS and CSM, we need to help both groups advance their goals.

References

National Council of Teachers of Mathematics. (1989). *Curriculum and evaluation standards for school mathematics*. Reston, VA: Author.

National Council of Teachers of Mathematics. (1991). *Professional standards for teaching mathematics*. Reston, VA: Author.

Part III
Issue Papers

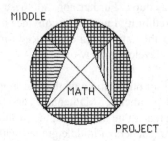

MIDDLE

MATH

PROJECT

The following papers were written by MIDDLE MATH participants on issues important in the preparation of middle grades teachers. The papers are organized into five main sections: methods; mathematics; diversity; assessment; and technology. The final paper of the section is an overall reflection on the experience of participating in this project.

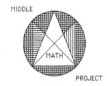
Issue Papers: Methods

Transforming the Investigative Lesson Model Into Middle Grades Classrooms

Ann Crawford and Donald Watson
University of North Carolina at Wilmington

Pedagogy in the mathematics classroom should be driven by dominant theories of learning. From the behaviorist and information processing theories, a content-centered approach to lesson design has given the preservice teacher a guide for planning and conducting instruction. However, with a constructivist learning base, the teacher is now asked to focus upon mathematics as problem centered and process oriented. If the goal of each lesson is student understanding — connecting and applying mathematical concepts rather than a set of procedural steps — a question then arises: What elements of the direct teaching model from Gagne's nine events of instruction need revision to correspond to a constructivist classroom more effectively (Gagne, Briggs, & Wager, 1972/1992; Gagne & Driscoll, 1988)?

The purpose of this paper is to discuss instructional events appropriate to a constructivist lesson plan. This is followed by a lesson plan guide for an investigative lesson that highlights these changes. Piaget's cognitive developmental theory (Piaget, 1953, 1964) and the Learning Cycle (Karplus & Thier, 1967) provide theoretical background for the investigative lesson guide. Finally, suggestions for transforming the investigative lesson guide into the middle grades classroom will be given.

Modifying the Nine Events of Instruction

In the first phase of Gagne's nine events of instruction, the teacher gains the attention of the students, provides the objective as an advanced organizer, and reviews prerequisite skills. In a constructivist lesson, a teacher still needs to gain students' attention and to motivate the lesson. Necessary prerequisite skills that apply in the investigation also need to be reviewed. However, if students are to build their own knowledge of the concepts through problem solving, an objective detailing the learning outcome is not needed. Rather, a general statement as an advanced organizer, such as "today, we will investigate...," can organize students' thoughts toward the problem solving.

In the second phase of the nine events of instruction, the teacher presents the stimulus material, provides learning guidance, and elicits performance of the learning outcome from the student. With a constructivist approach, rather than a teacher-presented stimulus, "the problem" is the stimulus for investigation by the student, who then builds meaning into the concepts. The necessary task of the teacher is first to explain the problem investigation, then elicit the performance through the investigation, and provide learning guidance while interacting with the students. Often the learning guidance from the teacher is minimal, with student interaction serving as the monitoring element. Thus, learning guidance and eliciting performance are reversed in the constructivist plan. The "learning task" replaces the "teacher's set of steps."

In the final phase of Gagne's nine events of instruction, the teacher provides feedback to the students, assesses the learning outcome, and provides for retention and transfer. With a constructivist lesson, the students themselves formalize the concepts that they are developing through sharing and discussing. The teacher will assess students' understanding during the class interactions. This learning outcome of "understanding" is quite different from learning outcomes defined as specific behaviors. The teacher may also utilize other methods of assessment, such as written explanations or journal reflections. The lesson can conclude with an extension of the concepts for retention and transfer. This final phase of a constructivist lesson includes assessment and summarizing the lesson, but the summary is accomplished by the formalizing of the concepts by students themselves.

A Constructivist Guide for Lesson Planning

In order to guide the design of a lesson plan, preservice middle grades teachers can be given the following outline.

Organization of a Lesson Plan for Investigative Lessons

I. *Motivating the Learning.* The teacher introduces the investigative problem and reviews any prerequisite skills that students will apply. A general objective as an advanced organizer may be given.

II. *Understanding the Problem.* The teacher needs to provide enough information and direction to the students so that students will understand the essential elements of the problem situation. Assessment of student understanding of the problem is needed before going on.

III. *Student Investigation.* As students investigate the problem, they hypothesize, synthesize information, and draw conclusions. They apply mathematical skills within the problem-solving situation. Students may be working in groups, sharing ideas, and coming up with a group solution. The teacher assesses the concept development, as students explore the problem, by questioning students and providing feedback to individuals or to groups. Preplanned questions may be used as a guide to examine understanding and solution methods.

IV. *Formalizing Concepts by the Students.* The teacher asks students to report their findings. Students share how they arrived at their conclusions. If students are working in groups, the group results are shared. Dialogue between students, discussing their reasoning, is a goal. The teacher continues to assess concept development.

V. *Closure and Extension of Concepts.* The teacher involves students in summarizing their new concepts. The teacher may want to ask the students to summarize the concepts orally, to write a reflective summary of what they have learned, or give a summary application. If the concept is to be continued the next class period, the teacher may want to wait to have students summarize their findings.

VI. *Homework.* To extend the problem for further development, the teacher can have students come up with their own new problems, or apply what they have learned. Homework may also involve introducing the investigation topic for the next class period.

Theoretical Foundations

The learning cycle (Karplus & Thier, 1967) has been suggested as a
model for an investigative approach to mathematics (Fleener, Westbrook, & Rogers, 1995; Juraschek, 1983). This model has sound
grounding in Piaget's cognitive development theory (Piaget, 1953,
1964). According to Piaget, learning is an active process and, for
middle grades students, based upon concrete experiences. Students
must formulate their own questions, problems, and hypothesis through
exploration with materials. This will cause disequilibrium as students
encounter new concepts that do not "fit" their present mental structures.
Students make sense of the information through accommodation as
they modify old concepts and create new mental structures. This new
information is reorganized to form new structures of thought.

The learning cycle (Karplus & Thier, 1967) involves three phases
based on Piaget's three processes of assimilation, accommodation, and
organization. The first phase of the cycle, *Exploration*, involves students interacting with a problem situation and beginning the assimilation process. As new data is assimilated and mismatches occur, disequilibrium results. In the *Invention* phase, students discuss, analyze
class data, and negotiate a common summary. Students are able to
make sense of their information through accommodating their thinking
to the data. In the final phase, *Expansion*, students generalize and apply
the concepts to their expanded experiences to continue to build related
concepts.

The learning cycle describes the necessary three phases for the internal
conditions of learning within a constructivist lesson, while the lesson
plan guide for an investigative lesson (above) provides the external
events of the "teacher-student interaction." The lesson plan guide,
designed from the same three phases of instruction, applies a constructivist model of learning derived from Piaget.

A new middle grades project that employs a similar three phase model
for investigative teaching is the *Connected Mathematics Project*
(Lappan et al., 1996). Lessons are designed using a guide: launch,
explore, and summarize. The launch phase includes introducing the
problem situation and is similar to the "explore phase" of the learning
cycle and similar to "motivate the learning" and "understand the
problem" in the constructivist guide to planning. The explore phase is
comparable to the "invention phase" and "student investigation" in the
previous models. Likewise, the summarize phase resembles the "ex-

pansion phase" of the learning cycle and "student formalizing of concepts" and "closure and extension" in lesson guide (above). This shortened form of a three-phase lesson guide can further benefit the preservice teachers' concept development of a constructivist lesson.

From the Lesson Guide to the Classroom

Mathematics reform in the 1960s also emphasized understanding and discovering mathematics, rather than rote memory of skills. However, these new pedagogical methods were not modeled for teachers, nor was their application into actual classrooms adequately supported (Davis, 1990). Lacking was an emphasis on learning theories to support the methods, because the prominent theory at the time was behaviorism. After millions of dollars were spent during the sixties, teaching methods in mathematics subsequently remained the same (Weiss, 1978, 1987). To integrate *investigative teaching* into the middle grades classroom, teacher educators must provide a collaborative effort to assure it is infused into all aspects of the middle grades preservice curriculum. Rather than lecturing on how to implement an investigative approach, preservice teachers must be allowed to explore, participate in, practice, and refine the investigative method in real teaching settings.

Middle grades preservice teachers often begin their induction into the classroom with a beginning instructional design course. These courses may be taught as generalist courses not specific to mathematics education. Mathematics educators must assure that models other than the direct instruction model (specifically, the investigative model) are included in this course. Such a course can introduce investigative teaching as based upon the cognitive developmental theory of Piaget and the learning cycle of Karplus and Thier.

Instruction in mathematical content courses and middle grades specific education courses can incorporate the investigative lesson guide. With such modeling, students can develop critical thinking, gain understanding of mathematical concepts they will teach, and internalize the investigative instructional model. Without this modeling, preservice teachers lack opportunities to assimilate new definitions for teaching mathematics into their cognitive structures of "mathematics teaching as a demonstration of a set of rules." The investigative model is also an effective instructional method for fostering critical thinking in the general education courses (Meyers, 1988).

The induction of the model in the middle grades classroom must include the preservice teacher practicing the model, receiving feedback, and making adjustments to refine teaching skills. This should be done within a variety of contexts. For example, in a methods course the students can be required to complete a case study of a middle grades student by participating in a 6-8 week tutoring field experience. The students can practice applying the three phases of instruction—launch, explore and summarize—by working individually with a middle grades student. The preservice teacher can keep a weekly journal reflecting upon the learning patterns of the student. The tutoring experiences involve sequenced lesson planning, instruction, and assessment of the student. A final part of the project can be a letter written to the parents describing the learning progress of the student.

In the methods class, microteaching a unit on middle grades geometry provides an excellent model for investigating lessons. Students in the methods class are divided into microteaching teams of six students. The same six investigations for 20-30 minute microteaching lessons are given to each team. The team divides the investigations so one student teaches each lesson to the peer group. For planning the lesson, students from different teaching teams that each have the same investigation can meet to share ideas about how to develop the lesson. The lessons are then presented by each student to the remaining five members of the team at a time scheduled with the professor. Besides the feedback provided by the professor, the team can also provide written feedback to the "student instructor." In addition to experience with lesson development and instruction, students participate in six different investigative lessons for middle grades geometry. Students can share copies of their lesson plans with team members.

Finally, preservice teachers need to experience planning, teaching, and assessment of investigative instruction within a middle grades classroom. This can first be done with a one lesson teaching experience during a six-week observation field experience in a middle grades mathematics classroom in conjunction with the methods course. The students can be assigned to observe five classes, then plan and prepare a short investigative lesson. Later in the student teaching practicum, the student can become more proficient with investigative instruction through continued experience in the classroom. Students can be required to submit a videotape of an investigative lesson as part of the grading requirements of the practicum.

Conclusion

The primary purpose of this paper has been to discuss how middle grades mathematics education programs can foster lesson development based upon a constructivist theory of learning. A lesson plan guide for an investigative lesson was presented and supported by discussion of the learning processes derived from the cognitive development theory of Piaget and the learning cycle of Karplus and Thier. Most importantly, suggestions were provided detailing how to infuse the pedagogy into middle grades preservice programs. By implementing the previously described experiences, middle grades preservice teachers can build pedagogical methods that support student reasoning and understanding.

References

Davis, R. B. (1990). Discovery learning and constructivism. In R. B. Davis, C. A. Maher, & N. Noddings (Eds.), *Constructivist views in the teaching and learning of mathematics* (pp. 93-106). Reston, VA: National Council of Teachers of Mathematics.

Fleener, M. J., Westbrook, S. L., & Rogers, L. N. (1995). Learning cycles for mathematics. *Journal of Mathematical Behavior, 14*, 437-442.

Gagne, R. M., Briggs, L. J., & Wager, W. W. (1972/1992). *Principles of instructional design.* New York: Harcourt Brace Javanovich.

Gagne, R. M., & Driscoll, M. P. (1988). *Essentials of learning for instruction.* Englewood Cliffs, NJ: Prentice Hall.

Juraschek, W. (1983). Piaget and middle school mathematics. *School Science and Mathematics, 83*(1), 4-13.

Karplus, R., & Thier, H. D. (1967). *A new look at elementary school science.* Chicago: Rand McNally.

Lappan G., Fey, J. T., Fitzgerald, W. M., Friel, S. N., & Phillips, E. D. (1996). *Connected Mathematics Project.* Palo Alto, CA: Dale Seymour.

Meyers, C. A. (1988). *Teaching students to think critically.* San Francisco: Jossey-Bass.

Piaget, J. (1953). *The origin of intelligence in the child.* London: Routledge & Kegan Paul.

Piaget, J. (1964). Cognitive development in children. *Journal of Research in Science Teaching, 2*, 176-186.

Weiss, I. R. (1978). *Report of the 1977 National Survey of Science, Mathematics, and Social Studies Education*. Durham, NC: Research Triangle Institute, Center for Educational Studies. (ERIC Document Reproduction Service No. ED 152 565)

Weiss, I. R. (1987). *Report of the 1985-1986 National Survey of Science and Mathematics Education*. Durham, NC: Research Triangle Institute, Center for Educational Studies. (ERIC Document Reproduction Service No. ED 292 620)

Use of Math Manipulatives to Encourage a Mathematically Literate Society

Ita Kilbride
Lees-McRae College

The mathematician George Polya (1980) stresses "the beautiful things" in the mathematics curriculum, not only information but know-how, independence, originality, and creativity. In this paper, I discuss "the beautiful things" elementary and middle grades student teachers explored using math manipulative materials, both commercial and homemade. I comment on how these materials contribute to the preparation of a work force for the 21st century which is prepared to absorb new ideas, perceive patterns, and solve unconventional problems.

The National Council of Teachers of Mathematics' *Curriculum and Evaluation Standards for School Mathematics* (NCTM, 1989) identifies five broad goals required to meet students' mathematical needs for the 21st century:

- *To value mathematics.* Students' experiences in school must bring them to believe that mathematics has value to them.

- *To develop confidence.* Students' experiences in school must be so numerous and so varied that they learn to trust their own mathematical thinking.

- *To solve problems.* The development of each student's ability to analyze and select appropriate problem-solving strategies that work individually or in conjunction with others is essential if students are going to be productive citizens.

- *To communicate mathematics.* Learning to read, write, and speak about mathematical constructions teaches the student to use the signs, symbols, and terms of mathematics.

- *To reason mathematically.* Students' experiences in schools must foster a climate where making conjectures, gathering evidence, making constructions, or building an argument are a natural extension of their math experience.

These five broad goals are addressed when elementary and middle grades math teachers encourage the use of commercial and teacher or student-made manipulative materials in their classrooms. There is mounting evidence that students, like adults, must construct and organize their mathematical knowledge before they can construct new or more refined relationships and understandings (Troutman & Lichtenberg, 1995, p. 25). The development, selection, adoption, and evaluation of math manipulative materials go a long way to meet these goals.

In fall 1996, several grade K-8 field experience students developed their own math/science manipulative materials as part of their portfolios for student teaching. When students and teachers construct their own materials reflecting their unique personalities and styles, they also construct their own pedagogical awareness. This, in turn, enhances their learning, reduces math anxiety, and stresses the "beautiful things" in mathematics curriculum. The students' reflective journals and comments in class reflected these outcomes. Certain dominant responses to the construction and use of math manipulatives emerged and are listed below:

- We learn a given subject by thinking, writing, reading, and constructing.

- Language and thought (including mathematical) are intimately related.

- Mathematical thinking becomes clearer as we seek to demonstrate it with manipulatives.

- We solved problems that we did not think we were capable of solving.

- It was hard to come up with the ideas at first, but it was great when it was done.

It was obvious from the enthusiasm with which the students presented, videotaped, and discussed their projects as part of an evaluation process that their assessment reflected the reform vision of school mathematics. The students learned from the activities and the construction of math manipulative materials. This, in turn, fostered growth towards high expectations in an approach that is philosophically in step with the mathematical needs of the 21st century. At least, this is the hope.

References

National Council of Teachers of Mathematics. (1989). *Curriculum and evaluation standards for school mathematics*. Reston, VA: Author.

Polya, G. (1980). On solving mathematical problems in high school. In S. Krulik (Ed.), *Problem solving in school mathematics, 1980 yearbook* (pp. 1-2). Reston, VA: National Council of Teachers of Mathematics.

Troutman, A., & Lichtenberg, B. (1995). *Mathematics, a good beginning: Strategies for teaching children* (5th ed.). Pacific Grove, CA: Brooks/Cole Publishing Co.

Integrating Mathematics and Methods

Susan Beal
Saint Xavier University

This paper challenges the movement towards general mathematics for elementary school teachers. During the fall of 1990, over 2400 mathematics educators at the university level, associated with the National Council of Teachers of Mathematics, were queried as to their awareness of the current mathematics education reform publications, and how the information from publications was integrated into the mathematics methods classes. Approximately 25% (56 out of 221 responses) of the faculty indicated that they taught an integrated mathematics and methods course for preparing elementary school teachers.

As states move toward more general mathematics requirements for elementary school teachers, this percentage should drop. I am sorry to see this trend. Integrated mathematics and methods courses tend to be rich in content, rich in their applications to classroom teaching, and meaningful in their approach to the learning of mathematics. Preservice teachers with a desire to have a concentration or minor in mathematics can pursue mathematics courses in addition to those that address the mathematics learned/taught in the elementary school curriculum. I agree that elementary school teachers desperately need more mathematics. But I disagree that changing the requirement of what mathematics is basic to teaching in an elementary school is the means to turn teachers-in-training into mathematics teachers.

In contrast to the separation of methods from the mathematics content courses taken by prospective elementary school teachers, I propose to discuss a two-semester integrated mathematics content and methods course for preservice elementary school teachers taught for the past twenty years at Saint Xavier University. What makes this course innovative is that the integration is accomplished by presenting material in a problem-solving form in a laboratory context—with preservice teachers working with manipulative materials as part of a cooperative group. Such a format allows prospective elementary school teachers to (re)learn mathematics and learn about the teaching of mathematics in a way they have not been exposed to previously (and in a way which is consistent with the NCTM *Curriculum and Evaluation Standards*, 1989). Thus, it sets a compelling paradigm for similar courses at other institutions to examine and emulate. At the very least, it will start, I hope, a dialogue about the desirability of integrating mathematics and teaching methods into one course.

General Course Description

The course under discussion is conducted in a typical laboratory model. Prospective teachers sit in groups of three, four, or possibly five around a table. In the cupboards surrounding the room are all the materials available to solve problems or answer questions posed in the laboratory manual, by the instructor or another class participant. These materials include Cuisenaire rods, counters, base ten blocks, weights, balances, meter sticks, textbook series from various companies, geoboards, etc., as well as pairs of scissors, paper clips, rubber bands, tape, and so on. Although some materials are suggested for use in the manual, pre-service teachers are free to use whatever materials they want. They are aware of the problems they will be investigating during any particular day, and are expected to have at least read the appropriate materials prior to the class meeting. At times, there are preparations that must be done at home, such as reading assignments or hands-on projects. For example, the prospective teachers construct a meter measure using ten decimeter lengths of construction cardboard and use it to find at least six objects: two objects that are approximately ten centimeters long, two objects that are a meter long, and two objects that have an area of a square decimeter.

The laboratory manual used in the courses was written for teachers-in-training and teachers-in-service to explore actively the mathematical concepts presented to students in the elementary school. The mathematical content explored over the two semester sequence is fairly standard: numeration systems and bases; operations with whole numbers, integers, and rational numbers; ratios, proportions, and percents; number theory; measurement and geometry; probability and statistics; and functions. Problem solving is infused throughout the curriculum. The courses are each three credit hours, but meet four hours each week because of the amount of material and the approach taken.

Teachers are expected to participate in their own learning. The participants are involved in an activity-oriented mathematics laboratory where they can explore mathematical ideas and learn the meanings and properties of the fundamental arithmetic operations. It is assumed that they will work in small groups in a cooperative atmosphere. They are expected to help each other, not only within their own group, but also acting as resource people for other groups. Explaining an idea to others helps clarify the idea for the explainer. The learning of mathematics is not a spectator sport, but a participatory one. It has been shown that small group cooperative learning is effective with children and adult learners. It is also hoped that the teacher will adopt small group learning in his/her own classroom.

As previously stated, a variety of concrete materials are used to facilitate the learning of mathematical ideas. The materials are embodiments of the mathematical ideas. Teachers use these materials to *play* with concepts, helping them construct their own mathematical knowledge. I believe that knowledge cannot be presented to individuals; it is something they must construct to own. A variety of appropriate materials are used in the laboratory activities to expose the mathematical ideas being considered and to emphasize that learning is unique to the learner. Using multiple materials and approaches to the same concept will enrich the classroom environment for all students regardless of level.

To emphasize that mathematics is not limited to school learning and has direct meaning in our lives, teachers are asked to read newspapers and newsweeklies to find stories about the practical use of mathematics. The underlying assumption for these exercises is that teachers must see the relevance of mathematics to help uncover it for their students.

Teachers also write a journal to reflect upon their own learning of mathematics, using a computer and submitting a disc at least once a month. Thus, the journal allows for a conversation between the instructor and the teacher. Learning to become a reflective practitioner takes time, and these courses are just a first step toward this goal.

The laboratory manual integrates mathematics and methods. Mathematical ideas needed for the development of an elementary school topic are introduced and developed using the appropriate concrete and pictorial models. This allows the user to (re)learn the mathematics in a non-threatening manner. It also introduces teachers-in-training and teachers-in-service to materials that will help elementary school students to learn mathematics. Over the course of the two semesters, teachers are expected to develop lesson plans, summarize and "critique" articles from mathematics journals, interview elementary school students, develop a graphing project, participate in a clinical experience, and analyze a textbook series on a particular topic.

It is not intended that the laboratory manual stands alone. A text that emphasizes the teaching/learning of mathematics helps put activities from the manual into perspective, offering background readings of various theories about how children learn mathematics, the construction of lessons and some activities and exercises to extend laboratory experiences to elementary school students. Textbooks in use in the elementary schools are needed for many of the experiences in the manual. These materials provide opportunities for exploring the mathematics taught throughout the grades, possible sequencing within mathematical topics, and comparisons between approaches used in various textbooks, as well as adherence to the vision of school mathematics presented by the National Council of Teachers of Mathematics. Readings in professional mathematics journals, such as *Arithmetic Teacher*, *Teaching Children Mathematics*, *Mathematics Teaching in the Middle School*, *Mathematics Teacher*, and *School Science and Mathematics*, are a necessary ingredient for professional growth. They enable the reader to put mathematical ideas into the context of the classroom, reinforce the idea that there is no one right way to teach mathematics, and allow for a continuation of dialogue of mathematical ideas and teaching strategies.

The role of the instructor varies throughout the course. At times the instructor acts as a resource person, helping teachers answer their own questions and encouraging them to *play* with mathematical ideas. Sometimes the instructor might lecture or encourage a class discussion, or help summarize and/or generalize some of the ideas that were *bantered* about during the class.

For example, in a number theory unit, while working on divisibility, the question arose as to how you know if a number is divisible by 11. A single digit multiple is obvious, but how can other multiples be explained. Teachers at different tables took responsibility for different multiples: three digits, four digits, and five digits, looking for patterns to help determine a generalization. The teachers exchanged work to determine if their generalizations could be applied to new data. The class then worked together to determine a generalization. However, the generalization was not *working* with numbers like 902 or 8250. We then briefly talked about modular arithmetic to help tie the discussion together and iron out the wrinkles that tend to bubble up during an inductive exploration.

Sample Assignments

Most of the teachers-in-training do not have an idea of what students in the elementary school know or understand. One of the first assignments given to the preservice teachers is to interview five elementary school students and ask questions related to estimating costs, measuring, number sense, attitudes toward mathematics, reading and writing numerals, counting, magnitude, mental calculations, students' approaches for solving standard textbook exercises and *word problems*. Results from these interviews (which are analyzed and written up by the teachers) are referenced throughout the two-semester course. For example, when asking the elementary school students to place 1/2 on a number line, third grade students usually will place 1/2 between the one and the two on the number line. These results are discussed during the unit on rational numbers, which takes place during the second semester.

A culminating activity to the functions, graphing, and statistics unit is a graphing project. Teachers choose a question which they want to answer or a position they would like to take; devise a data-collection experiment which would shed light on the question or support or refute the position; and collect, organize, and present the findings to the class. They also discuss how they would adapt the activity for use in an elementary classroom at two distinct grade levels, how they would introduce it to the students, the need for presenting the information in a graph, and questions they might ask students in each of the grades about the graphs, as well as possible follow-up activities.

Another activity that I enjoy is called "Adopt a Solid." The teachers are asked to choose a solid from the classroom's supply of models. They are given two weeks to find out as much as they can about the solid. They should be able to describe it in many ways. For example, how many faces does it have and what shapes are they? How many vertices (corners) does it have? How many edges? Etc. Teachers are asked to make a model of the solid, and find out what role it plays in the real world, its uses, etc. To present this information to the class, teachers write an essay or poem about the solid, explaining why the solid is special. The presentations are usually at the end of the second semester, and are a wonderful way to end the year.

Summary

Contrast the richness of this integration of mathematics with the push to have prospective elementary teachers separate the mathematics they learn from the methods they learn to teach mathematics. It will never be possible to include all that prospective teachers need to see and do, either in mathematics or in the methods of teaching mathematics. But what we can offer is a model by which teachers-in-training can learn and feel successful in their learning, work towards becoming problem solvers, and hopefully learn to enjoy (I want to say love) mathematics.

References

National Council of Teachers of Mathematics. (1989). *Curriculum and evaluation standards for school mathematics.* Reston, VA: Author.

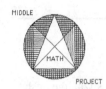
Risk-Taking in the Classroom

Jennifer Earles Szydlik
University of Wisconsin-Oshkosh

New curriculum materials call for a transformed pedagogy. Teachers
are expected to have students work on problems and facilitate discussion on solution methods and content. Instructors must guide classroom
conversation, comprehend and value a variety of solution methods,
choose mathematical ideas to pursue, and offer counter-examples to
student claims when appropriate. Additionally, the teacher is asked to
substitute problems for worksheets, class discussions for the carefully
crafted lecture, consistency and logic for mathematical authority, and
flexibility and risk for certainty. This approach is intimidating and risky
for most teachers because it makes mathematics problematic and
messy, and it can reveal their own mathematical weaknesses. Students
are sure to become confused and even frustrated before they see a
solution, asking questions the teacher has not anticipated, and creating
mathematics and arguments the teacher does not know. This new
method of teaching requires strong content knowledge, confidence, and
the belief that mathematics is something that can be figured out—that
consistency and logic will ultimately prevail.

Consider the following descriptions of two fictional middle grades
classrooms. In the first classroom, students are working cooperatively
in small groups to find all the semi-regular tessellations of a plane.
They have boxes of regular polygons and are sketching each different
pattern they find. Two groups have abandoned the manipulatives in
favor of adding up the angle measures to see which combinations of
angle measures of the polygons sum to 360 degrees. Two students are
having a mathematical debate about whether one of the conjectured
patterns can be built. Another group is measuring the angles of the
various polygons, and everyone is getting a different answer. Yet
another group believes that they are done with the problem and are
experimenting with their polygons as they wait for the group discussion
to begin. The teacher, Ms. Jones, is moving among the groups asking

questions, helping students focus on mathematics, and gathering information for a conversation about angle measure, regular polygons, and tessellations.

In the second classroom, the students are working in cooperative groups on the same problem. The room is loud and unruly. Some students are playing with the manipulatives, while others are frustrated and have stopped working on the problem. Students are arguing about an answer and the teacher, Mr. Smith, is trying to mediate—however, he is not sure which student is really correct. Another group is taking measurements in such a sloppy manner that they are not collecting any useful information. Mr. Smith desperately is trying to keep the students on task and to put out mathematical fires raging all around the room. He hopes that no colleague or parents drop by to see this lesson and is wondering how he can bring order to mathematical chaos.

The first classroom provides ample opportunity for mathematical conversation. The class can discuss solution strategies, accuracy in measurement, ways of figuring out the angle measure of a regular polygon without measuring, distinctions between patterns of polygons that will fit around one vertex but cannot be extended to the plane and those which do tessellate the plane, and how to make an argument that all the possible tessellations have been found. The second classroom does not provide any significant mathematical opportunity. Mr. Smith must spend his time addressing misconceptions and teaching students how to measure angles (again). He must go home tonight and research why some polygons seem to tessellate based on their angle measures but cannot actually be constructed and present that to the class tomorrow (at the expense of tomorrow's lesson plan).

What is the difference between these two classrooms? Why does Ms. Jones believe this is a successful lesson and Mr. Smith thinks it's a failure? The answer is this: the success or failure did not depend on *what happened* in each class—exactly the same events could have occurred in both rooms. The success or failure occurred in the *interpretation* of the events; where Ms. Jones saw mathematical learning, Mr. Smith saw chaos; where Ms. Jones saw excitement and debate, Mr. Smith saw frustration and an argument that required mediation; where Ms. Jones saw an opening for learning, Mr. Smith saw misconceptions, his failure to teach prior material, and confirmation of his fear of being an inadequate teacher. (I am not claiming that Mr. Smith *is* an inadequate teacher, just that he may see himself as such.)

It is time to acknowledge that transformed teaching can be terrifying. Mr. Smith risked wasting class time working on a problem that many students would not solve. He risked the class not being capable of distinguishing between reasonable strategies and solutions and those which are mathematically bankrupt. He risked the realization or confirmation that he did not fully understand the problem. He risked feeling incompetent. He risked having his students become frustrated or unruly. He risked his colleagues believing that his students were not learning.

What are the differences between Ms. Jones and Mr. Smith that prompted them to interpret events in their classrooms so differently? There are several possibilities. Perhaps Mr. Smith was less experienced with the teaching method, perceived resistance from students or colleagues, had weaker content knowledge, or had beliefs about what mathematics is or how mathematics is done which are inconsistent with the pedagogy.

Of all these possible explanations for the differences in perception, the last seems more serious. Suppose Ms. Jones believes that mathematics is logical and consistent and something that human beings can figure out, and this allows her to see her classroom as a hotbed of mathematical activity. This is what mathematics is supposed to look like; students are supposed to experiment, make mistakes, and clarify their misconceptions. Mathematics will be created from this process. Ms. Jones has confidence that the class will solve the problem at hand and that she and her students will learn from the lesson.

On the other hand, suppose Mr. Smith believes that mathematics is a collection of facts and procedures to be learned and applied appropriately. He is the mathematical authority in the classroom and believes that he needs to know the answers to his students' questions. He has few tools for creating mathematics or confidence that he and his students can know a mathematical answer without looking it up in a textbook or appealing to a higher authority. For him, having students experiment and create mathematics *makes no sense*.

Thompson (1984) identified three different views of mathematics which teachers hold and discussed how each impacts how teachers perceive instruction:

- The *problem solving* view in which mathematics is described as a "… continually expanding field of human creation and invention, in which patterns are generated and distilled into knowledge,"

- The *Platonist's* view in which mathematics is thought to be "… a static but unified body of knowledge, a crystalline realm of inter-connecting structures and truths, bound together by filaments of logic and meaning," and

- The *instrumentalist* view in which mathematics is described as an "… accumulation of facts, rules, and skills to be used by trained arti-sans skillfully in the pursuance of some external end" (Thompson (1984, p. 10)).

Thompson (1984) found that teachers (like the fictional Ms. Jones) who held a problem-solving view of mathematics were more likely to employ activities that allowed students to construct mathematical ideas for themselves, whereas teachers (like Mr. Smith) who held an instru-mentalist view of mathematics were more likely to teach in a prescrip-tive manner emphasizing rules and procedures. (For a synthesis of this literature see Thompson, 1992.)

For teacher educators, there are at least three possible strategies for addressing the "failure" of the second classroom: 1) We can acknowl-edge that not all teachers will be comfortable using the new curriculum materials and accept that many teachers will adopt more traditional curriculum materials or teach the new materials in a traditional way. 2) We can work to change teachers' beliefs about the nature of mathemat-ics. 3) We can support and encourage risk-taking. I assert that we must employ all three strategies.

The charge will not be an easy one. Educators committed to the trans-formation of teaching will have a difficult time accepting that there exist people who should be traditional teachers — traditional pedagogy is consistent with their beliefs about mathematics. For them, teaching in a transformed manner is illogical and too risky. To make matters worse, in general, attempts to alter preservice (and inservice) teacher beliefs have not been successful (Lerman, 1987; Schram & Wilcox, 1988; Shirk, 1973), and the mechanism by which beliefs are altered is not understood. Risk-taking is also not well understood.

The research literature does provide some assistance in the realm of changing teacher beliefs. Both Collier (1972) and Meyerson (1978) observed a slight change in the beliefs of preservice elementary teachers from the view of mathematics as rigid and the role of the teacher as the demonstrator of facts, to mathematics as creative and the role of the teacher as supporting the students in learning the material. Meyerson attributed this success to being able to generate doubt in the minds of teachers. This doubt arose in problem-posing situations that created confusion and controversy suggesting that we need to model for our students the transformed pedagogy—not only to help them learn the content, but also to help them see mathematics as creative and logical.

How do we promote risk-taking in the classroom? Here are two ideas. First, we must acknowledge that the transformed pedagogy is terrifying for many teachers and understand that what looks like an exciting mathematics lesson to us may look like an ugly mess to our preservice teachers or their colleagues. Second, we must take risks in our own classroom and make the process transparent to our undergraduate students. Consider the possibility that by employing a transformed pedagogy, we are not only modeling teaching but also risk-taking itself. Conveniently, as many of our prospective teachers do not want us to generate doubt and create confusion and controversy in the classroom, doing so will provide us an abundance of risk-taking opportunities.

References

Collier, C. P. (1972). Prospective elementary teachers' intensity and ambivalence of beliefs about mathematics and mathematics instruction. *Journal for Research in Mathematics Education, 3*, 155-163.

Lerman, S. (1987). Investigations: Where to now? In P. Ernest (Ed.), *Teaching and Learning Mathematics, Part 1 (Perspectives 33)* (pp. 47-56). Exeter, England: University of Exeter School of Education.

Meyerson, L. N. (1978). Conception of knowledge in mathematics: Interaction with and applications to a teaching methods course (Doctoral dissertation, State University of New York, 1978). *Dissertation Abstracts International, 38*, 733A.

Schram, P., & Wilcox, S. K. (1988). Changing pre-service teachers' conceptions of mathematics learning. In M. J. Behr, C. B. Lacampagne, & M. W. Wheeler (Eds.), *PME-NA: Proceedings of the tenth annual meeting* (pp. 349-355). DeKalb, IL: Northern University.

Shirk, G. B. (1973). *An examination of conceptual frameworks of beginning mathematics teachers*. Unpublished doctoral dissertation, University of Illinois at Urbana-Champaign.

Thompson, A. G. (1992). Teachers' beliefs and conceptions: A synthesis of the research. In D. A. Grouws (Ed.), *Handbook for research on mathematics teaching and learning* (pp. 127-146). New York: Macmillan.

Thompson A. G. (1984). The relationship of teachers' conceptions of mathematics teaching to instructional practice. *Educational Studies in Mathematics, 15*, 105-127.

From Preparation to Practice: Teachers' Dilemmas, Tensions, and Frustrations

Beverly Ferrucci
Keene State College

*At times I feel that it is so difficult teaching in middle school.
Many of the teachers in this building feel that they are very
good teachers (and they are) but in social studies and language arts—not in mathematics. They didn't learn enough
mathematics in college, and so they only teach computation,
computation, and more computation. They also think that just
because they are using manipulatives or having their students
work in groups that they are doing a great job teaching
mathematics to middle school students. It's so frustrating, it
just makes me want to scream!*

These feelings, expressed by an experienced middle school teacher,
point out two very important concerns for educators—inadequate
preservice mathematical preparation of K-8 teachers and inappropriate
use of manipulatives and other pedagogical techniques. These topics
were found to be prevalent themes among a number of middle school
teachers who I interviewed as an assignment for the MIDDLE MATH
Conference.

In many instances, the middle school teachers expressed a great deal of
dissatisfaction with their undergraduate mathematical preparation,
which they attributed as a major cause of the inadequacy in their
mathematics background and knowledge. As one teacher stated,

*It's easy to learn effective teaching techniques during your
teaching career. Colleagues are always willing to share good
lessons, and you can get many ideas by attending conferences
and workshops. You can't always do that, though, with the
mathematics. So, it is imperative that you study as much
mathematics content with good professors before you graduate with your teaching degree. Otherwise, you may never be
able to find a colleague who can explain the mathematics to
you. I would have preferred much more instruction in mathematics during my undergraduate program.*

One teacher also expressed a desire to have been given more mathematical instruction and less methods-type courses before graduating
from her baccalaureate program:

> *The problem, as I see it, is that we are given many "cute"*
> *middle school mathematics methods courses. They do an ade-*
> *quate job preparing you how to teach, but they are much too*
> *weak in exposing you to upper grades mathematics. Many of*
> *the teachers in my school only feel comfortable with their*
> *mathematical knowledge up to pre-algebra.*

Embedded in this statement is yet another dilemma for middle school
teachers—the notion that an understanding of higher-level mathematics
is not needed. In fact, several of the teachers felt that an understanding
of any mathematics beyond algebra is not necessary for a middle school
teacher. As two seventh grade teachers commented,

> *Why do I need to know and understand concepts from algebra*
> *and above? I'm never going to be teaching it.*

> *How do you expect me to teach critical thinking skills when I*
> *have never studied them myself? Learning the basics—*
> *computation and drill work—that's all I needed and it worked*
> *well for me. So, that's the way I am going to teach math in my*
> *middle school classes.*

Many middle school teachers believe teaching abstract mathematical
concepts falls under the domain of high school teachers. They view
their role as being the teacher who reinforces and ensures students
master basic skills before proceeding to high school to study "real"
mathematics. A sixth grade teacher summarized her thoughts:

> *If students memorize and learn their basic facts, then critical*
> *thinking skills will develop by themselves later. That's for the*
> *high school teachers to do in their mathematics classes. We*
> *are just supposed to make sure the students know their facts*
> *and can perform their calculations before they get to high*
> *school.*

Another area of contention was the inappropriate use of manipulatives
and other pedagogical techniques. Many of the interviewed teachers
discussed the fact that merely using manipulatives or other methodolo-

gies in the classroom does not ensure that quality mathematics learning is taking place. One teacher voiced her frustration:

> *For instance, last week I visited the other sixth grade mathematics class. The students were using Cuisenaire rods, but not for learning fractions or anything like that. Their assignment was to see how many letters of the alphabet they could build using just five dark green rods. Just where was the mathematics learning? Was this a worthwhile way to spend a 55-minute class? I don't think so!*

However, not all teachers feel this way. One teacher pointed out that after reading articles about the educational reform movement and the poor performance of American students on mathematical tests, she has changed her philosophy and method of teaching:

> *I'm not saying that I'm the greatest mathematics teacher. In fact, I used to think just like the other teachers in my school. However, I began to see that my students could perform computational problems very quickly, but without possessing a deep understanding of the concepts and certainly without an understanding of the whole picture. That made me change my way of thinking and my way of teaching. Now I try to incorporate the computational work with problem-solving skills and integrated activities.*

Another teacher described his attempts to change his mode of teaching:

> *Now I've become more open-minded and willing to try new ideas to help improve my teaching. Originally, my style was to write on the board, lecture to my students, and drill them with computational problems. Later, I moved to writing on the overhead. Now I make the students write on the overhead and explain their methods and solutions to the class. This process allows for a great deal of mathematical discussion, and it also involves much, much more than just computation.*

Although several factors may be responsible for a feeling of mathematical inadequacy in many of these middle school teachers, one of the main reasons is the quantity of mathematical experiences preservice K-8 teachers receive in their college instruction. In many teacher education programs in colleges and universities in the United States, only one or two courses in mathematics are required. These courses may often only involve a discussion of topics found in the elementary and middle school curriculum or a presentation of ways to introduce these topics. Very seldom do these courses cover topics from higher-level mathematics.

In order to teach mathematics effectively, middle school teachers need to acquire an understanding of mathematical topics that extend beyond those they will be teaching. They need to be given opportunities in their college instruction to explore more advanced mathematical topics and to develop critical thinking and problem solving skills, resulting in the development of confidence and competency to teach mathematics. They must also be made aware of the multitude of materials, activities, and opportunities available to enrich their mathematical knowledge. Perhaps one way to achieve this goal is to encourage teachers who are innovative in their teaching (and who do possess a firm understanding of mathematical concepts) to become role models and leaders in their individual schools, districts, and mathematics teacher organizations, sharing their knowledge and teaching skills with other middle school teachers.

Mathematics for Preparing Middle Grades Mathematics Teachers

Susann Mathews
Wright State University

Having participated in the MIDDLE MATH Project, I have developed the following proposal for the mathematics that should be included in the preparation of teachers of middle grades mathematics. While recognizing that the teaching of mathematics is as important as the content, the majority of this paper discusses the content of the mathematics itself. We must realize that if mathematics is to become part of a dual major of mathematics and science for middle school preservice teachers, it must contain mathematics appropriate to a mathematics/science degree within a College of Science and Mathematics, and it must be consistent with the mathematics content described in *A Call for Change* (Leitzel, 1991), *Curriculum and Evaluation Standards for School Mathematics* (NCTM, 1989), and *Professional Standards for Teaching Mathematics* (NCTM, 1991). It is my hope that the material presented here meets that need.

Themes that should run throughout the curriculum, as well as specific courses and topics, must be addressed. To help prospective teachers teach the new middle grades mathematics curricula, the mathematics courses should all have an element of reflection about the mathematics the teachers are learning. Each course should help teachers develop flexibility in their thinking. Teachers should be learning by discovery with the instructor as facilitator. All the courses should be infused with appropriate technology, and curriculum should develop sound teacher beliefs. Furthermore, after problems and activities, the following questions should be posed to help the teachers learn to recognize the mathematics and its big ideas:

- What was the mathematics?

- Why would we want to learn this?

- Are the mathematical ideas important ones? Why?

These "experiences provide the core from which teachers will eventually build learning environments for their own students" (NCTM, 1991, p. 128).

A strong pre-calculus course should begin the program. It should focus on the concept of function. Linear, quadratic, and general polynomial functions, rational functions, and exponential and logarithmic functions are important examples that should be introduced and connected through applications.

Early in the program, just after pre-calculus, teachers should take three (four quarter-credit-hour) courses in calculus. These classes teach the concepts in a traditional two-quarter sequence of calculus, but in a manner consistent with the methodology recommended in the *Curriculum and Evaluation Standards for School Mathematics*, including more inquiry, problem solving, and writing. The calculus courses should consider diverse applications, should be concept driven rather than manipulation driven, and should make connections to algebra, geometry, and the middle school curriculum. Teachers in the middle grades should have the background necessary to prepare students for all secondary mathematics courses, including calculus. Furthermore, if these courses are studied early in the program, the science courses that are taken as part of a dual mathematics/science major can be calculus-based. A yearlong sequence provides a chance for teachers to study a mathematical topic in depth and gain exposure to other ways to solve problems. Courses required after the calculus sequence can then assume that teachers already have a calculus background and can be taught at the 300-level and above.

The following courses should be required after calculus (each taken for four quarter-credit hours): a two-sequence course analogous to *Fundamental Mathematical Concepts* (often known as *Mathematics for Elementary School Teachers*), a probability and statistics course, a geometry course, an algebra and functions course, and a capstone course in problem solving and mathematical modeling.

The two-sequence course analogous to *Fundamental Mathematical Concepts* should contain number concepts and relationships, number theory, discrete mathematics, some geometry and measurement, and the algebraic structure of our number system. These courses go into more depth and breadth of topics than typical mathematics for elementary teachers courses.

A probability and statistics course should be included because skills in these areas are necessary for fundamental mathematical literacy. In this course, teachers collect data based on experiments or surveys, organize and interpret the data, and formulate convincing arguments based on their data analyses. In addition to furthering teachers' understanding of data analysis and statistics, these experiences further their problem-solving skills. There should be a component in which the incorrect use of statistics is studied by analyzing and critiquing arguments that are based on incorrect statistical use. Because teachers often lack correct intuitive notions of probability, they need to make connections between physical probability problems and symbolic representations of these problems. "Probability instruction must compete with [middle school students'] possibly strongly held intuitive beliefs and strategies that may be inconsistent with instruction" (Bright and Hoeffner, 1993, p. 87). Therefore, throughout their study of probability as students, future middle grades mathematics teachers should be asked to explain their reasoning in order to openly examine their intuitions so that later, as teachers, they can be aware of their own conceptions and ideas. Teachers should also plan and conduct experiments and simulations to determine relative frequencies and compare those results to their study of theoretical probabilities. In this course, teachers learn to analyze and interpret statistical information and to make decisions in the presence of uncertainty. Appropriate statistical and probability software should be used to help meet this goal.

A geometry course should be included in which both Euclidean and non-Euclidean geometries are studied. Teachers should have concrete experiences with geometric figures and relationships prior to formal axiomatic study. Thus, they can develop their own geometric under-standing and learn about the stages through which geometric under-standing evolves (Leitzel, 1991). Teachers will have had concrete geometric experiences in the courses analogous to *Mathematics for Elementary Teachers* and should now be ready to build on those experiences. The course created for this purpose should include trans-formation geometry, as well as coordinate and synthetic geometry, and should help teachers make connections among these three systems. Teachers should investigate properties and relationships of shape, size, and symmetry in two and three dimensions. This understanding can be extended by including a study of spherical geometry, using the Lénárt Sphere and appropriate technology.

An algebra and functions course should be part of the mathematics curriculum. In this course, teachers examine the ideas of patterns, variables, and functions from a more sophisticated viewpoint than they did previously in pre-calculus. The emphasis is on making connections within mathematics, between mathematics and the real world, and between the mathematics they are studying and the middle grades mathematics they will eventually teach. To further this understanding, multiple representations of relationships and functions should be stressed: physical models, charts, graphs, numerical approaches, and equations and inequalities to describe real-world relationships. Graphing calculators and software are particularly appropriate here.

A capstone course in problem solving and mathematical modeling should conclude the required mathematics content. As mathematics teachers, one of our weakest characteristics is our lack of ability to demonstrate connections between mathematics and the real world. A major thrust of new middle-grades mathematics curricula is to develop students' understanding of connections within and outside of mathematics. The problem-solving and mathematical modeling course will strengthen preservice teachers' understanding of connections. Furthermore, it can help them realize that real mathematics problems take more than ten minutes to solve, that false starts are common, that persistence pays off, and that problems often have a variety of different approaches and different answers depending on initial assumptions and interpretations. Finally, mathematical modeling can expose the preservice teacher to changes in the nature of mathematics, the way we do mathematics resulting from the availability of technology, and provide them with the self-confidence necessary to help their future students become problem solvers.

References

Bright, G. W., & Hoeffner, K. (1993). Measurement, probability, statistics, and graphing. In D. T. Owens (Ed.), *Research ideas for the classroom: Middle grades mathematics* (pp. 78-98). New York: Macmillan.

Leitzel, J. R. C. (Ed.). (1991). *A call for change: Recommendations for the mathematical preparation of teachers of mathematics*. Washington, DC: The Mathematical Association of America.

National Council of Teachers of Mathematics. (1989). *Curriculum and evaluation standards for school mathematics*. Reston, VA: Author.

National Council of Teachers of Mathematics. (1991). *Professional standards for teaching mathematics*. Reston, VA: Author.

Subject-Specific Frameworks in the Preparation of Middle School Mathematics Teachers

Anita Bowman
University of North Carolina – Greensboro

For the duration of the MIDDLE MATH Project, my thoughts have centered around the concern of how to define a well-connected, pre-service curriculum for middle grades mathematics teachers. In particular, I have considered ways to modify the four common NCTM (1989) mathematics curriculum standards—mathematics as problem solving, mathematics as communication, mathematics as reasoning, and mathematical connections—into an analogous set of mathematics teacher education curriculum standards—teaching and learning as problem-solving, teaching and learning as communication, teaching and learning as reasoning, and teaching and learning connections. I find it intriguing to play with the question, "How might a middle grades mathematics teacher education program be defined if based on the four teaching and learning standards listed?" It seems to me that underlying such a curriculum definition must be a belief that "good" teaching and learning is (a) globally identifiable in practice and (b) content specific in nature.

What do I mean by "globally identifiable in practice?" Consider a patchwork quilt done in a "Lone Star" pattern. The quilt may display an 8-pointed star with each point of the star composed of 64 rhombuses, and the remainder of the quilt composed of a series of squares, triangles, and rectangles—all pieces of assorted colors and textures but artistically joined to produce a pattern giving the sense of "whole cloth." That is, when we see the finished quilt, we do not see hundreds of individual pieces of cloth of specific colors and shapes, but rather a single pattern that we proclaim to be "beautiful."

As mathematics teacher educators, we sometimes experience an analogous vision as we observe in classrooms. In such instances we see students engaged in problem solving. We observe their solution strategies and understand their reasoning as they communicate with one another, often aided by skillful questioning by a teacher. We see students comparing their solutions and solution strategies, identifying similarities and differences in approach, and connecting concepts one with another. Like the quilt, we proclaim the pattern we see in the classroom to be "beautiful" without truly attending to the components that make this pattern possible.

Just as we recognize the beauty in the single picture produced by the artistic arrangement of hundreds of pieces of fabric, we recognize, globally, the beauty of a teaching/learning pattern that works. However, as teacher educators, our task is not simply the recognition of good teaching and learning in classrooms. We must somehow abstract the essential elements of good teaching/learning instances and transmit them to our preservice teachers. It is with these last two tasks that we struggle. Then again, maybe not. Maybe we have invested most of our energies into the task of transmitting important information to our preservice teachers without adequately thinking through and determining what the essential elements are. Maybe this last task requires so much of our effort precisely because we have not adequately identified the essential elements we want to transmit.

Perhaps we have major problems determining what the essential elements are. It sometimes seems as if teacher education curricula have no more design than "crazy" quilts—quilts made by sewing randomly-ordered pieces of fabric together until the resulting quilt-top is of sufficient size to cover a bed. That is, for the middle grades mathematics preservice teachers we might define the curriculum as a collection of mathematics, psychology, pedagogy, and general education courses, without a sense of any underlying framework that serves to connect the courses into an outcome-based whole. Perhaps this idea can be clarified by contrasting examples.

Consider an undergraduate curriculum leading to a degree in chemistry. Typically, such a curriculum will incorporate courses in inorganic, organic, physical, and analytical chemistry. These courses are held together by a somewhat simple framework—the electronic structure of atoms. Once a student understands the electronic structure of atoms and the relationship of electronic structure to the periodic table, the student has achieved the essential understanding needed to develop understanding of chemical bonding, reaction mechanisms, material properties, thermodynamics and kinetics, to name a few topics. That is, the electronic structure of atoms provides an essential underlying framework upon which much of the chemistry curriculum may be based.

A somewhat analogous situation is found in undergraduate mathematics curricula. The concept of function provides a framework for a large proportion of the content defined in mathematics curricula. That is, a robust understanding of mathematical function through an understanding of representations of function (graphs, tables, algebraic formulas, verbal descriptions, and situations) and translations among representations of functions provides a student with an underlying framework upon which much of the mathematics curriculum may be based.

Now ask yourself the question, "Is there an analogous framework that might serve to provide a base for middle grades mathematics teacher education curricula?" I think there is—and I think this framework must be content specific in nature.

Before I go any further with this line of thinking, I want to highlight something that is beginning to happen in some K–3 mathematics teacher education programs. Efforts labeled as Cognitively Guided Instruction (CGI) began with inservice K–3 mathematics teachers and, more recently, have been extended to preservice programs. CGI is based on content-specific knowledge and relies heavily on the standards of problem solving, communication, reasoning, and connections. Teachers learn to differentiate among problem types and use a combination of research-base and personal experience with students to diagnose student progress toward content understanding. CGI teachers then base instructional decisions on very detailed content-specific information gathered on students within a specific class. As teachers learn more about ways students learn specific content, they begin to use this knowledge more in making instructional decisions. In so doing, the teacher truly assumes a professional role of diagnosis, prescription, and instructional creativity.

Is there an analogous content-specific framework that might have similar results in the preparation of middle grades mathematics teachers? That is, it seems important to ask if there might be a connecting framework that can span the preservice curriculum, connecting learning mathematics with teaching mathematics, research in children's learning of mathematics, children's developmental growth, and curriculum design. It seems to me that the concept of function, as mentioned above as a framework for undergraduate mathematics curricula, might be expandable into a framework for middle grades mathematics teacher education curricula.

The research base on students' understanding of mathematical function, at all grade levels, is extensive. This research base lends itself to reinterpretation within the pentagonal model of function understanding—

a model that details function understanding by focusing on representations (graph, table, algebraic formula, verbal description, and situation) and translations among representations. The pentagonal model might then be used to help preservice teachers learn about student learning — as evidenced in hundreds of research articles.

Furthermore, the pentagonal model might be used to provide a framework for mathematics content courses in the preservice curriculum — spanning theoretical work as well as applications, in general, and mathematical modeling, in particular. Within the preservice curriculum teachers might be explicitly taught the model and assigned tasks requiring reflection on their own thinking and understanding within the framework of the model. Then, within the preservice curriculum (as teachers progress to pedagogy and curriculum components) they might reflect on how curriculum tasks and instructional procedures connect to the pentagonal model. Thus, the pentagonal model might provide a powerful framework for connecting content across the preservice curriculum.

In conclusion, what I am proposing is that within a general framework of four standards — teaching and learning as problem-solving, teaching and learning as communication, teaching and learning as reasoning, and teaching and learning connections — and a content-specific framework provided by the pentagonal model, middle grades mathematics teacher education curricula may be re-designed to provide much needed continuity among courses and to help preservice teachers develop important reflective, diagnostic, and prescriptive skills needed to create the patterns we all recognize as "beautiful" examples of classroom teaching and learning.

References

National Council of Teachers of Mathematics. (1989). *Curriculum and evaluation standards for school mathematics*. Reston, VA: Author.

Middle School Education Programs: How Specialized Should They Be?

Judy Curran Buck   **William E. Geeslin**
Plymouth State College   University of New Hampshire

Many of the presentations at the first conference of the MIDDLE MATH Project emphasized that specialized courses for future middle school mathematics teachers are needed due to the special characteristics of adolescents and make-up of middle schools. It is somewhat curious that such interests in specific middle school training have arisen only recently. This may be partially the result of the transition in nomenclature and pedagogy involved in the change from "junior high schools" to "middle schools." Nonetheless, an immediate question arises: which preservice courses should be directed entirely toward middle school mathematics teachers?

Before attempting a partial answer to this question, we introduce some constraints. Not all colleges and universities have sufficient numbers of education students to allow for the entire mathematics education program to be directed at one school level. Even in a large university, specialized courses must be offered less frequently, and hence, may cause scheduling problems for students. College admissions officers claim most students change majors at least once during their academic career. Students entering mathematics or mathematics education are often uncertain about whether or not they wish to teach, as well as undecided about at which level they wish to teach. Interactions among students in mathematics and mathematics education are desirable as well as interaction among students intending to teach at different grade levels or in different content areas. Finally, institutions have general education requirements and general teacher certification requirements that limit the number of courses one can reasonably require in a middle school mathematics major. So, most institutions must carefully consider which courses are applicable to all mathematics majors and which must be directed toward a specific group.

Because the authors teach at relatively small institutions and are interested in maintaining as flexible a program for as many students as possible, focusing a course on middle school mathematics teaching is appropriate only when there is a compelling pedagogical reason to do so. For instance, we do not find much reason for separating students out in introductory educational psychology courses, feeling that middle school preservice teachers would benefit from understanding the psychological changes that occur over the course of the K-12 years. We take as an axiom that these students should complete as many regular college-level mathematics courses as is possible given various institutional and certification constraints. Students preparing to teach should not be enrolled in a different calculus course than students majoring in mathematics or engineering. However, there are compelling reasons to separate students by level and content area in methods courses. The curricula differ across school levels and the most effective methods of presenting these curricula differ. Thus, we would recommend at least one specialized one-semester methods course for middle school mathematics teachers (specialized both in content and pedagogy).

Most model programs noted at the MIDDLE MATH Project meeting suggested a series of integrated content and methods courses for middle school teachers. First, we believe all future middle school mathematics teachers should, at a minimum, complete the first two years of the same mathematics courses as mathematics majors, including calculus, some form of linear algebra, and introduction to mathematical proof. A computer science course involving programming is strongly recommended as well. Prior to the "methods" course, we see a need for two specialized courses (three semesters): (1) a two-semester course in number, geometry, and algebra; and (2) a one-semester course in probability, statistics, and discrete mathematics. Methods and content should be integrated throughout these courses. Course one might enroll elementary mathematics education specialists as well (but not general elementary majors who have not had calculus). Course two could easily include secondary (high school) majors. We suggest this series culminate in a "methods" course, which would extend the previous pedagogical knowledge gained through special courses and field experiences, and focus primarily on psychology, pedagogy, assessment, and teaching practice.

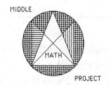

Preparing Mathematics Teachers for Culturally-Responsive Teaching

Catherine Gardner and Dan Swetman
Mercer University

The cultural diversity of our society is increasing, and the middle school classroom is a reflection of that change. The mono-cultural classroom of the 1950s and 60s no longer exists. Teacher preparation programs must evolve to prepare preservice teachers to teach to a diverse community of learners. The National Council of Teachers of Mathematics (NCTM) is "committed to the principle that groups underrepresented in mathematics-based fields of study should be full participants in all aspects of mathematics education" (*NCTM Handbook*, 1996-97 p. 19). Underrepresented groups are those who choose not to take advanced mathematics courses or enter professions involving mathematics. These groups consist of Hispanics, African-Americans, females, Native Americans, Alaskan Natives, and Native Pacific Islanders. Mathematics instruction must reflect and appeal to *all* students.

Mathematical learning and the culture of underrepresented groups are linked. Stigler and Baranes (1989) refer to this combination as "a new sociology of mathematics [that] has arisen that takes its premise that the foundations of mathematics are to be found through the examination of the cultural practices in which the activities of mathematics are embedded" (p. 359). The cultural influences on mathematical understanding are real. These influences are manifested through "cultural tools, cultural practices, and cultural institutions" (Lee & Slaughter-Defoe, 1995 p. 359). Ethnomathematics as a concept is accepted as a framework influencing instruction (Ascher, 1991; Frankenstein, 1990; Stiff & Harvey, 1988; Zaslavsky, 1979, 1993).

Intertwining mathematical concepts with culture can be accomplished using several strategies. Teachers must design curriculum and instruction around real-life experiences of culturally diverse students (see for example, Anderson, 1990; Frankenstein, 1990; Joseph, 1987; and Lee & Slaughter-Defoe, 1995). Tate (1993) and Secada (1993) recommend that mathematics instruction for culturally diverse students be grounded in their everyday lives and daily struggles. Mathematics instruction should surpass rhetoric and empower the students to improve or change their lives (Tate, 1993).

Learning style theory has been recognized as an essential component of successful teaching (Silva & Strong, 1995). The conceptual basis of learning styles comes from the theory that students react to their educational experiences with consistent behavior and performance patterns. These patterns consist of a mixture of cognitive, affective, and physiological behaviors originated and maintained by culture, personality, and brain chemistry (American Association of School Administrators, 1991; Bennett, 1990). Research on learning styles of culturally-diverse students is not conclusive, but it does indicate significant possibilities for improving academic achievement for culturally-diverse student populations (Irvine & York, 1995).

Silva and Strong (1995) developed the *Diverse Teaching Strategies for Heterogeneous Student Populations* curriculum to aid teachers in planning and teaching the culturally-diverse population in schools today. They used the *Myers-Briggs Type Indicator* (MBTI) as the theoretical basis for their research on learning styles. The fact that children and adults have preferred learning styles has been well documented (Dunn & Dunn, 1978; McCarthy, 1987). Learning style theory recognizes the diversity in children's construction of knowledge. Although the terminology varies, the planning and teaching of culturally diverse populations should also be linked to Jung's Psychological types (Jung, 1923). Learners can be grouped as *sensory-thinker*, *sensory-feeler*, *intuitive-thinker*, and *intuitive-feeler*. Students usually have at least one preferred learning style.

- Students whose dominant learning style is *sensory-thinker* want the teacher to tell what they need to know and how they need to master the information to complete the work and make a grade of 100. Such students excel in listing, describing, or recalling information.

- The learning style represented by the *intuitive-thinker* student is a desire to understand the *why* and *how* behind the academic content. This student learns best when allowed to explain a concept, test a hypothesis, or solve a dilemma.

- The student who is identified as *sensory-feeler* has greatest success when working with others on content that has personal connections. These students excel when involved in cooperative learning activities, which identify and use their feelings.

- The *intuitive-feeler* student learns best when using activities, which allow artistic expression, divergent thinking, or making unusual connections. This student experiences greatest success when allowed to design unique products and use art or music to create ideas or original solutions to problems.

Silva and Strong (1995) stress that most instruction and assessment is centered around the *sensory-thinker* learning style. The least taught and valued learning style is *intuitive-feeler*—the group of learners with the highest high school dropout rate. To address this inequity, Silva and Strong developed a teaching strategy they called *task rotation*. In *task rotation*, an academically challenging concept is chosen, then is taught and assessed in each learning style. Each learning style is given equal value and time. When children are allowed to succeed in their preferred learning style, the dropout rate is reduced.

Teacher-preparation programs provide models for how to teach culturally diverse student populations, incorporating strategies such as learning styles theories and real-life, relevant mathematical context. The challenge for mathematics teacher education programs is to develop strategies to prepare the preservice teacher with the theory, skills, and aptitude to teach *all* students.

The five National Science Foundation–funded *standards-based* middle school curriculum projects, detailed earlier in this monograph, provide excellent examples of academically-challenging mathematics presented in a manner that involves *all* students. *MathScape: Seeing and Thinking Mathematically*, for example, has *ethnomathematics* as an underlying basic principle. *The Language of Numbers* unit begins with the idea of numbers as one of humanity's greatest inventions. Students study number systems based upon the contributions of different cultures. Teaching to a wide range of learning styles abounds in the various activities. Patterns grow from a rod-painting activity, a shepherd's field, a calendar, and an old man's will. Activities such as enlarging patterns, and analyzing how or why a pattern develops encourages learning for *sensory-thinker* students. The *intuitive-thinker* student uses patterns to describe a personal experience and is allowed to create a pattern that has never before existed. These patterns can be used to reflect upon the everyday experiences of students.

The *Middle-school Mathematics through Applications Project* uses activities which start with a slice of the real world. The project then creates a scenario involving a problem or issue. To solve the problem or provocative issue, students must use a significant amount of mathematics. In the *Antarctica Project,* students must design a home base in which they can live for six months. They must use their cultural experiences to predict what changes are necessary for survival. Learning-style theory is emphasized as to what problems they choose to solve and how they process the problem. Assessment in each learning style is an option.

These projects are only examples of a quality curriculum, which is being prepared for middle school students to learn and middle school teachers to teach. How can teacher preparation programs prepare teacher candidates and inservice teachers to teach these new programs? Teacher education programs are in a unique situation. Most of the preservice teachers have been in a classroom for approximately 14,000 hours before entering college. With all their experiences, most pre-service teachers enter their educational programs with strong convictions about how a classroom should be run. They are not *blank slates* onto which professors write knowledge. For preservice and inservice teachers to accept new theories and pedagogy, they must undergo cognitive disequilibrium (Scheurman, 1996). One successful method of introducing this disequilibrium is through the *Reflective Teaching Cycle* (Karen Shultz, personal communication, November, 1996). The teacher and an associate plan a lesson using culturally responsive teaching strategies. Then they videotape the implementation of the lesson in a middle school classroom. After teaching and videotaping, the videotape is viewed and reflected upon. Invariably, the teachers observe teaching and learning practices they want to improve. This activity provides the disequilibrium and, frequently, teacher behavior changes to incorporate culturally responsive pedagogy.

Middle school mathematics teachers (and the college programs which produce these teachers) must evolve to meet the needs of a changing clientele. Mathematics curriculum developers are meeting the challenge through NCTM-based standards which incorporate learning styles strategies, real-life meaningful problems, and culturally-rich learning environments. Schools of Education must respond by modeling effective teaching behaviors in education classes (as well as mathematics classes) using reflection as a method of improving instruction, and introducing and using the *Reflective Teaching Cycle* to improve instruction.

References

American Association of School Administrators. (1991). *Learning styles: Putting research and common sense into practice*. Arlington, VA: Author.

Anderson, S. (1990). World math curriculum: Fighting Eurocentrism in mathematics. *Journal of Negro Education, 59,* 348-359.

Ascher, M. (1991). *Ethnomathematics: A multicultural view of mathematical ideas*. Pacific Grove, CA: Brooks/Cole.

Bennett, C. I. (1990). *Comprehensive multicultural education (2nd ed.)*. Boston: Allyn & Bacon.

Dunn, R., & Dunn, K. (1978). *Teaching students through their individual learning styles*. Reston, VA: Reston Publications.

Frankenstein, M. (1990). Incorporating race, gender, and class issues into a critical mathematical literacy curriculum. *Journal of Negro Education, 59,* 336-351.

Irvine, J. J., & York, D. E. (1995). Learning styles and culturally diverse students: A literature review. In J. Banks & C. M. Banks (Eds.), *Handbook of research on multicultural education* (pp. 484-497). New York: Macmillan.

Joseph, G. (1987). Foundations of Eurocentrism in mathematics. *Race and Class, 27,* 13-28.

Jung, C. G. (1923). *Psychological types*. (H. G. Baynes, Trans.). New York: Harcourt, Brace & Co., Inc.

Lee, C. D., & Slaughter-Defoe, D. T. (1995). Historical and sociological influences on African American education. In J. Banks & C. M. Banks (Eds.), *Handbook of research on multicultural education* (pp. 348-371). New York: Macmillan.

McCarthy, B. (1987). The 4MAT System: *Teaching to learning styles*. Barrington, IL.: Excel.

National Council of Teachers of Mathematics. (1997). *1996-97 Handbook: NCTM goals, leaders, and positions*. Reston, VA: Author.

Scheurman, G. (1996). Philosophical chairs: A technique to elicit prior knowledge and beliefs. *Newsletter for Educational Psychologists.* American Psychological Association.

Secada, W. G. (1993). *Towards a consciously-multicultural mathematics curriculum.* Paper presented at the Teachers College Conference on Urban Education, Teachers College, Columbia University, New York.

Silva, H. F., & Strong, R. (1995). *Teaching and learning strategies for heterogeneous classrooms. The principles and practices of thoughtful education.* Princeton Junction, NJ: Hanson, Silva, & Strong.

Stigler, J., & Baranes, R. (1989). Culture and mathematics learning. In E. Z. Rothkopf (Ed.), *Review of research in education.* (Vol. 15, pp. 253-307). Washington, DC: American Educational Research Association.

Stiff, L., & Harvey, W. (1988). On the education of Black children in mathematics. *Journal of Black Studies, 19,* 190-203.

Tate, W. F. (1993, April). *Can America have a colorblind national assessment in mathematics?* Paper presented at the annual meeting of the American Educational Research Association, Atlanta.

Zaslavsky, C. (1979). *Africa counts: Number and patterns in African culture.* Chicago: Lawrence Hill Books.

Zaslavsky, C. (1993). *Multicultural mathematics: Interdisciplinary cooperative-learning* activities. Portland, ME: J. Weston Walch.

Difficulties of Reform
in Small Rural Schools

Judy Curran Buck   **William E. Geeslin**
Plymouth State College   University of New Hampshire

The National Council of Teachers of Mathematics (NCTM) and the
National Science Foundation (NSF), through projects such as the
MIDDLE MATH Project at East Carolina University, have been
promoting changes in the training of middle school mathematics
teachers. The funded efforts (in particular) often seem to require,
explicitly or implicitly, that reform projects deal with large numbers of
teachers and students. For example, NSF guidelines suggest certain
costs be tied to a certain number of teachers as an ideal level of support,
and reviewers criticize proposals that exceed these amounts. Obviously,
it is more cost efficient to aim efforts at large groups of more than half
a dozen teachers and more than 100 students. However, there exist
numerous locales that are not large, ethnically diverse urban/suburban
areas, but still are in great need of funding for improvement and revision of their mathematics education programs. Such schools still suffer
from having uncertified or under-certified mathematics teachers,
insufficient public funding, and educationally disadvantaged students.
Our purpose here is to outline some of the difficulties in attempting to
improve these programs and suggest one possible solution.

The state of New Hampshire has been attempting to implement revisions in the middle school mathematics programs through a variety of
means, including teacher training programs at the three state institutions of higher education: the University of New Hampshire, Plymouth
State College, and Keene State College. New Hampshire is a state of
approximately one million people, mostly comprised of small towns,
rural areas, and a few small cities. Public schools are funded primarily
through local property taxes with relatively little assistance from the
state. One consequence of little state assistance is little state control—
hence, conditions at schools are quite varied. New Hampshire students
enrolled in grades 5-9 may find themselves housed in a junior high
school, a middle school, an elementary school, or secondary school.
Their mathematics teachers may hold an elementary, middle, or secondary school certificate (with or without a specialization in mathematics). State regulations are such that, depending on the composition of
school levels in a particular school (7-12, 6-8, K-8, etc.), a school may

be in compliance with regulations even though students in grades 5-8 are not being taught mathematics by a teacher with a middle school mathematics certificate. Thus, resulting efforts to reform mathematics instruction and curricula must address a wide variety of institutions, along with a set of teachers who range in mathematics training from essentially none to those having a master's degree in mathematics.

In New Hampshire, preservice education, particularly at the university, must serve a significant number of out-of-state students. Although these students come primarily from the northeast, their home states have widely differing public education systems.

Inservice education in New Hampshire focuses primarily on New Hampshire, but many schools have only one mathematics teacher, making systemic reform a real challenge. Reform in small schools takes place only when the one or two mathematics teachers at the school are committed to reform. Likewise, in a small, usually underfunded school, administrators must become advocates of the reform mode of thinking. Many schools do not pay the tuition costs of their staff, nor do they have funds for substitutes, which would allow a teacher to be absent from school for professional purposes. At the same time, there are over 100 public "middle" schools, and college level instructors cannot visit all the schools during a school year—even on an annual basis. Teachers who cannot get release time can neither visit another school to observe nor demonstrate new classroom techniques.

Though New Hampshire has a very active state mathematics teachers' association, which provides workshops and sessions through conferences and inservice programs, not all teachers are allowed by their schools to attend. The college system and state department of education provide traveling "road shows" from time to time, but again, these programs can visit only a few schools for only a short period of time. The university system's efforts are largely unfunded but occasionally are able to obtain either Eisenhower or federal funds. To supplement this support, we suggest that federal and state agencies attempt to fund summer programs that provide mathematics content and pedagogy to middle school teachers and administrators. During the academic year, funds should go to provide travel for college level mathematics educators to make monthly visits to the middle school classrooms of these teachers. In addition, a few exemplary middle school teachers should be funded to visit schools to observe practices and demonstrate alternative teaching approaches. This process would take more than one year to be successful, but it would allow smaller schools to receive assistance in revising their programs. To reduce costs in schools where there are several middle school mathematics teachers, a local *teacher leader* could be chosen through summer programs. Providing teachers with appropriate levels of mathematical knowledge is probably the greatest challenge in this process—particularly for those teachers who do not attend the summer programs.

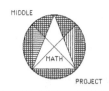

Issue Papers: Assessment

Issues Related to the Assessment of Pre-service Mathematics Teachers

Michaele F. Chappell and Denisse R. Thompson
University of South Florida

Assessment is a highly complex process with many factors influencing the decisions educators make when evaluating preservice mathematics teachers. Yet, assessment is an important component of teacher preparation. It provides mathematics educators with means for monitoring preservice teachers' growth in the knowledge and skills required to become effective instructors.

Recently, the mathematics education community has engaged in serious dialogue about the nature of assessment as applied to elementary, middle school, and high school students. Specifically, the *Assessment Standards for School Mathematics* (NCTM, 1995) elaborates on the use of assessment to help all students have better opportunities to learn mathematics and develop the mathematical tools expected of them. In addition to tests, teachers should engage students in other forms of assessment, such as projects, oral demonstrations or presentations, performance assessments, journals, and portfolios. As Cooney, Bell, Fisher-Cauble, and Sanchez (1996) note, these other forms of assessment can encourage students to think critically about mathematics and provide valuable insight to teachers when modifying instruction. Similarly, multiple forms of assessment impact preservice teachers by forcing them to think critically about mathematics content, pedagogy, and the interaction between these two elements when creating effective instruction. For teacher educators, multiple forms of assessment enable us to monitor our students' progress, evaluate their achievement, and make instructional decisions about our own practices (NCTM, 1995, p. 25).

This article highlights several pertinent issues regarding the assessment of preservice mathematics teachers. Though many of these issues have been communicated within the context of teaching K-12 school mathematics, they have yet to receive the same level of dialogue within the domain of teacher preparation. The issues raised here are not exhaustive; however, we believe they are critical aspects of discussions on the assessment of preservice mathematics teachers, in both pedagogy and content courses.

Issue: Modeling Assessment Practices

Preservice mathematics teachers themselves must engage in multiple forms of assessment and reflect on their experiences if we expect them to value the use of multiple forms of assessment when they enter K-12 classrooms. Reflecting on such assessments in mathematics content courses emphasizes that some aspects of mathematical power are best assessed through avenues other than timed, on-demand tests. For instance, projects provide an opportunity to work on a problem/situation over an extended period of time; open-ended or non-routine problems require time to reflect on the problem and/or try multiple strategies in seeking solutions. Reflecting on such assessments in content-pedagogy courses emphasizes that the instructional presentation determines the extent to which the content is readily understood. For example, a textbook critique forces preservice teachers to consider how the sequencing of topics and surrounding activities interact to facilitate student learning.

The issue of modeling assessment is quite critical. When confronted with the realities of teaching in the classroom, teachers often rely on the models of teaching they have seen and experienced. They quickly recall their former teachers' actions, their own actions, and very seldom retrieve information simply communicated to them. That is, beginning teachers tend to emulate our actions by *doing what we do, not just what we say!* Thus, if we only talk about multiple forms of assessment but fail to use them in our own courses, we send a subtle message about the extent to which we ourselves value these other forms of assessment.

Chappell and Thompson (1994) present several examples of assessments used in their mathematics pedagogy courses. These assessments allow students the opportunity to demonstrate their knowledge of the content learned in the courses, apply that knowledge, make reasonable steps toward becoming a professional, and understand the nuances involved in teaching mathematics. By reflecting on these experiences, preservice teachers are better positioned to generate, implement, and evaluate multiple forms of assessments in their own mathematics classrooms. Modeling multiple forms of assessment in mathematics courses for preservice teachers is certainly a viable way to promote the *Assessment Standards* in teacher preparation programs and enable preservice teachers to understand the vision of assessment as addressed by this document (NCTM, 1995).

Issue: Modifying Course Exams

One component of multiple forms of assessment is the use of course exams—in-class or take-home tests focusing on a wide range of topics. It is worthwhile for teacher educators to reflect on the design of such exams, ensuring they formulate questions and/or situations that embody and reflect the philosophical themes of the *Assessment Standards*. Although focused on middle-school and high-school mathematics courses, Thompson, Beckmann, and Senk (1997) suggest strategies for modifying test items to reflect the use of technology, the incorporation of applications, and the use of reasoning. These same strategies are applicable to post-secondary mathematics courses as educators attempt to broaden the range of questions presented on course exams. A change in a few exam items may significantly affect the overall tone of the exam.

A variety of question formats should also be featured in content-specific pedagogy courses. For instance, an exam may contain a combination of *short answer items* (e.g., listing, multiple-choice, matching, true-or-false, completion), *demonstration items* (open ended responses in which students have to carry out an algorithm or a procedure using hands-on materials; an explanation is usually requested), and *discussion items* (those that call for elaboration on a particular topic to a degree beyond simple justification or explanation). These varied item formats represent efforts toward understanding students' abilities to integrate and synthesize content and pedagogy.

Issue: Grading and Assessment

At the college level, assessment and grading are strongly interrelated. One issue facing teacher educators is determining the portion of a course grade that is based on written course exams and the portion that is based on other forms of assessment. Many students perform better on those assessments that are not in-class course exams. For instance, it is not uncommon for a student to earn a modest *B* on written course exams, perform in the *A* range on the other assessments required for the course, and receive an *A* as a course grade. Given this case, mathematics educators at times are concerned that too much of the course grade comes from non-exam assessments. This struggle may illuminate how mathematics educators themselves perceive the meaning of the *grade* that students receive.

However, mathematics educators should realize that not all mathematics learning can be assigned a grade. It could be argued that invaluable learning occurs for preservice teachers when they conduct non-exam assessments; this learning could more than make up for what they do not demonstrate on a course exam. For instance, when preservice teachers conduct a *Pupil Study* or *Spatial Activity Interview* with a small group of children (Chappell & Thompson, 1994), they often learn much more from the teaching situation than what they ultimately convey in their final summary reports. This knowledge clarifies, substantiates, or raises additional questions about their conceptions of certain mathematics content and how children learn that content from the instructional interaction that occurs between teachers and children. Hence, using multiple forms of assessments in preservice programs facilitates a broad range of learning opportunities, only some of which may be reflected in a course grade.

Closing Remarks

The issues presented above identify critical aspects challenging teacher educators as they attempt to implement changes in the assessment of preservice teachers. We have raised these issues not so much to provide answers, but to prompt thinking among those responsible for preparing middle grade teachers. Thinking about these issues promotes conversations and discussions that will be helpful in finding relevant solutions as mathematics educators teaching content and content-pedagogy courses mold and refine their own philosophy about what is important in becoming a mathematics teacher.

We invite the reader, if not already engaged in using multiple forms of assessment, to reflect on the issues presented in this paper and begin broadening their assessment base. We encourage the reader to begin with one assessment alternative in a course; as you become more comfortable with that alternative, you will be in a position to try another alternative as well. Over time, the means by which you assess preservice mathematics teachers will be significantly enriched.

final examination grades. In each instance, the instructor prepared several tasks of equivalent difficulty and importance for the student to complete. At the time of assessment, the student was randomly assigned a task and given five minutes to consider an approach as the task was completed. The student would then complete the task, giving explanations as the task was completed. The instructor, using a prescribed list of probing questions, would evaluate the student's work.

A typical task during the final conference was: "Use your grapher to display a complete graph of $y = 3x^3 + 2x^2 - 9x - 6$. Find the zeroes of the function correct to the nearest thousandth." The list of probing questions included:

- Is your graph complete? (How do you know?)

- How did you locate the zeroes? (Can you do it another way?)

- What is the value of the y-intercept? (Can you find it another way?)

- If $x = 11$, then what value(s) will y have? (Can you do it another way?)

- If $y = {}^-2$, then what value(s) will x have? (Can you do it another way?)

In each case, responses to the first question provided an assessment of the minimum acceptable performance and responses to the second question encouraged the student to demonstrate an understanding of the process. Students received a copy of their performance assessment at the end of the conference.

The third form of performance assessment employed in *Algebra Through Technology* was a project that involved the design of an experiment. The collection and analysis of data was an integral part of the entire course. Early in the semester, these experiments were carefully controlled in order to address specific algebraic and calculator skills. For example, one assignment involved the analysis of a year's utility information, such as average daily cost of gas or electricity, average monthly temperature, and average daily use of gas or electricity. The question under consideration was how to best predict your monthly utility costs (NCTM, 1995).

As the semester progressed, however, the experiments became more open-ended and required greater understanding of the practical aspects of the situation. For example, for a group project in the final month of the course, students were assigned an experimental situation and asked to produce a well-designed experiment that could be effectively used in a middle school algebra classroom. The original idea, *It's Simply Marbleous,* is from materials produced by the AIMS Education Foundation (AIMS, 1987). Each small group completed the original experiment that involved exploring the effect the slope of a ramp had on the distance a marble would roll. They then decided what other factors might affect the distance and designed a complete experiment which included the original experiment and their extension. The group was expected to devise a set of probing questions and an assessment model for their experiment. The final result was presented to the class in both written and oral form. Their classmates evaluated the experiment informally, making suggestions and asking for clarification, and the instructor completed the formal evaluation.

For preservice teachers, peer evaluation of their experiment was very informative. Viewing the situation from the perspective of a student and a teacher gave each of them simultaneous insights into the difficulties of clear communication in assignments and into the amount of effort required to produce a worthwhile performance assessment task.

Conclusion

An assessment program must provide information about each students' progress toward understanding mathematics. It must also help teachers make instructional decisions. The performance assessments described above accomplished these objectives efficiently and effectively. To paraphrase Gail Burrill (1997), president of the National Council of Teachers of Mathematics, in order to improve mathematics education, you must "show me the mathematics." And in order to evaluate students' progress, they must "show us the mathematics."

References

AIMS Education Foundation. (1987). *Math + science: A solution.* Author.

Ann Arbor Public Schools. (1993). *Alternative assessment.* Palo Alto, CA: Dale Seymour Publications.

Burrill, G. (1997). President's message. *NCTM News Bulletin, 4,* 3.

National Council of Teachers of Mathematics. (1989). *Curriculum and evaluation standards for school mathematics.* Reston, VA: Author.

National Council of Teachers of Mathematics. (1995). *Algebra in a technological world*. Reston, VA: Author.

Stenmark, J. K. (1991). *Mathematics assessment, myths, models, good questions, and practical suggestions*. Reston, VA: National Council of Teachers of Mathematics.

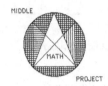
Issue Papers: Technology

The Effective Use of Technology in Mathematics Teacher Preparation

Zhonghong Jiang and Edwin McClintock
Florida International University

Advanced technology is increasingly pervasive in everyday life. More and more educators believe that the use of technology can effectively facilitate the teaching and learning of mathematics. This belief has been reinforced since the appearance of innovative technologies in mathematics education, including dynamic geometry software such as the *Geometer's Sketchpad* [GSP] (Jackiw, 1991), computer algebra software such as *Maple* and *Mathematica*, spreadsheet programs such as *Microsoft Excel*, graphing calculators such as the TI-83 and TI-92, and a variety of other powerful electronic tools. These technologies are highly interactive so that whenever a student's actions yield a reaction on the part of the machine, it in turn sets the stage for interpretation, reflection, and further action on the part of the student. Using these technologies, one can make powerful resources immediately available to aid thinking or problem solving, provide intelligent feedback or context-sensitive advice, actively link representation systems, and generally influence students' mathematical experience more deeply than ever before (Kaput and Thompson, 1994). In addition, the rapid computing speed of computers and graphing calculators can free students from tedious calculations and allow them to concentrate on conceptual understanding. By opening a new, colorful world to students, technology can greatly motivate students, stimulating their stronger interest in mathematics. Based on these considerations, the NCTM *Standards* (1989) for grades 5-8 emphasize the effective use of technology as one of the chief features of the reform curriculum. In recent years, many research studies (Edwards, 1991; Jiang, 1993; Manouchehri, 1994; Olive, 1991; Thompson, 1992; Thompson and Thompson, 1990) have provided evidence supporting the belief that students benefit from the use of technology.

Teachers are the most important factor in the use of technology in mathematics education. Without qualified teachers who are interested and have enough knowledge, experience and confidence in integrating technology into their mathematics teaching, students could not gain the benefits mentioned above. In order to become qualified teachers, our preservice instructors must receive adequate technology training in our teacher preparation programs and have an appropriate amount of practical experience in using technology with middle school students.

Technology in Teacher Preparation

Our position is that the effective use of technology should be a necessary and important component of our middle school mathematics teacher preparation programs. We should not only integrate the use of technology into all program courses wherever appropriate (especially mathematics content and methods courses) and involve it in students' field experiences, but also have at least one specific technology course which focuses on learning mathematics with technology (i.e., conducting technology-based sense making, problem solving, and mathematical reasoning) and exploring ways of teaching mathematics with technology.

There are several major reasons for us to have this position. First, in order for their future students to become strong mathematical problem solvers in technology-rich environments, the preservice teachers themselves should first be strong problem solvers. Second, they should be familiar with (or at least know the most important features of) different technologies, such as various software packages, graphing calculators, Calculator-based Laboratory (CBL), Multimedia, and Internet so they will be adept at choosing the right technology for best enhancing their future students' mathematical learning in any specific learning situation. Third, most inservice teachers know very little about the up-to-date technology, have little opportunity to consider a richer approach to their teaching, and lack necessary materials and methods for using technology as a teaching tool. Because most districts in the United States offer only approximately four days of release time per year for professional development, the inservice training is, generally speaking, limited. In addition, the one-day or two-day workshop training format can only give teachers a general sense of a specific technology. It is not sufficient to help teachers reach a degree of proficiency in using that technology. It is very important that there are technology experts among the teachers, who can offer long-term and persistent help when needed. In this sense, it is significant for our graduates to become sophisticated in using technology so that they can play the role of such experts (or more accurately, agents of change) after being assigned to schools. While being new teachers who need help with teaching strategies and ways to work with children, our graduates can effectively help inservice teachers with using technology. As a matter of fact, several new teachers who graduated from our program in recent years have been doing a great job as technology experts at the schools where they are working.

Without a systematic preparation in using technology for learning and teaching mathematics, it is not realistic to expect our preservice teachers to handle these tasks well when they take teaching positions at schools. It is also not realistic to provide the systematic technology training simply by using pieces of technology in different courses where instructors must consider more factors than the use of technology.

Carefully designed technology courses can serve the important purpose of substantially strengthening the mathematical background of pre-service teachers. Great opportunities are offered for them to revisit school mathematics, to construct deeper understanding of basic mathematical concepts through investigations and explorations, and hence to develop stronger problem solving and mathematical reasoning abilities. This is especially significant for states like Florida, where a legislation-fixed upper limit of 120 hours to graduation has depleted opportunities for enriched teacher training programs and almost no instruction time can be used to offer courses such as *Geometry for Middle School Teachers* and *Algebraic Concepts for Middle School Teachers*.

Two related issues are important to consider when designing a degree program involving a major emphasis on technology. First, different technologies offer different opportunities, so it is inappropriate to favor one technology over another. For example, it is inappropriate to use calculators (even graphing calculators) to the exclusion of computers. Computers offer many major benefits over calculators, including high-resolution graphics and strong animation features. Further, a computer-trained individual is educated in manners consistent with society's needs and the individual's need for changing job requirements and the diversity of occupations that an individual will encounter in his/her lifetime. Calculators are a substantively less useful tool, but have the advantage of being cost effective.

A second important issue is that technology can and should be learned with and in the context of another discipline or area of application. It is necessary for preservice teachers to be productive as members of society—having knowledge and skills in multiple disciplines or multiple application areas. Likewise, in preparing to teach mathematics in an interdisciplinary, connected way, prospective teachers should learn mathematics in an interdisciplinary, connected way.

Technology in Content and Methods Courses

The mathematics education faculty should work collaboratively with mathematicians to explore how to integrate appropriate technology into college mathematics courses. The initial step may include using computer algebra software such as *Maple* and graphing calculators such as the TI-85 or TI-92 in calculus courses. These technologies have been found by a number of mathematicians (Herod, 1996; Kenelly, 1996; Lopez, 1996) to be a great help in facilitating dynamic and interactive visualization for calculus concepts and problems, and in linking the visualization to the numerical and symbolic aspects of calculus to develop students' conceptual understanding. Likewise, we should consider using statistics software, such as *Datadesk,* and calculators with statistics components in statistics and probability courses.

Within the mathematics methods course, technologies should be used for investigations, applications, communications, problem solving, and as a suggested teaching tool. For example, we presented the following problem to our preservice teachers: "A traveler wishes to visit each of five cities in series, beginning and ending in the same city. However, the traveler does not wish to visit the same city more than once. What is the cheapest circuit of the cities that the traveler can make?" Students used various methods to analyze this problem, trying to understand it and solve it. Some of them chose to use technology. Figure 1 shows one of the solutions presented in a spreadsheet.

Leg	Price
Atlanta to Chicago	169
Atlanta to LA	319
Atlanta to Miami	198
Atlanta to NY	279
Chicago to LA	349
Chicago to Miami	329
Chicago to NY	179
LA to Miami	439
LA to NY	419
Miami to NY	249

This is the cheapest route. For reasons of economy, LA to Chicago appears as Chicago to LA. However, if the prices are changed, the route will be updated. Unfortunately, the prices must be kept at least very slightly unique for the text to update properly.

Atlanta to LA
Chicago to LA
Chicago to NY
Miami to NY
Atlanta to Miami
Total cost =   1294

The real route will nevertheless be found, and the total cost will reflect this.

The database below lists all 12 of the possible routes passing through each city once. Since every city is connected to every other, we begin with 5! permutations for visiting the cities. However, the starting city doesn't matter, so by choosing always to begin with Atlanta, we reduce the possibilities to 4!. But the direction of travel doesn't matter either, so there are finally only 12 possible routes.

Leg1	leg 2	leg 3	leg 4	leg 5	Total
169	349	439	249	279	1485
169	349	419	249	198	1384
169	329	439	419	279	1635
169	329	249	419	319	1485
169	179	419	439	198	1404
169	179	249	439	319	1355
319	349	329	249	279	1525
319	349	179	249	198	1294
319	439	329	179	279	1545
319	419	179	329	198	1444
198	329	349	419	279	1574
198	439	349	179	279	1444

Leg1	leg 2	leg 3	leg 4	leg 5	Total
319	349	179	249	198	1294

Price	Price	Price	Price	Price
319	349	179	249	198

These little "databases" were used as extract ranges for the cheapest route, and then to help convert the result into text.

Figure 1. One solution of the circuit problem in the methods class.

Specific Technology Course(s)

If possible, two specific technology courses should be designed and offered to the middle school preservice teachers. One of them could be named *Learning Mathematics with Technology* (treating the preservice teachers as learners of school mathematics in technology-rich environments). The other could be named *Teaching Middle Grades Mathematics with Technology* (concentrating on technology-based teaching ideas and strategies). If, for the time being, it is difficult to offer two technology courses, the availability of at least one such course is a must. In this situation, the course should cover both learning and teaching aspects mentioned above, and be designed around content strands recommended by the NCTM *Standards*. Because the middle grades mathematics that the preservice teachers are expected to teach is the direct foundation for and transition to high school level mathematics, a sound understanding of the two levels of school mathematics is necessary for all preservice teachers. Therefore, the course should deal with both educational levels and balance well between them.

The emphasis of this course should be on exploration of various mathematics contexts to learn mathematics, to pose problems and problem extensions, to solve problems, and to communicate mathematical demonstrations by using various software applications and graphing calculators. Students' classroom explorations, follow-up investigations, and projects that require the full range of mathematical final examination grades. In each instance, the instructor prepared several tasks of equivalent difficulty and importance for the student to complete. At the time of assessment, the student was randomly assigned a task and given five minutes to consider an approach as the task was completed. The student would then complete the task, giving explanations as the task was completed. The instructor, using a prescribed list of probing questions, would evaluate the student's work.

A typical task during the final conference was: "Use your grapher to display a complete graph of $y = 3x^3 + 2x^2 - 9x - 6$. Find the zeroes of the function correct to the nearest thousandth." The list of probing questions included:

- Is your graph complete? (How do you know?)

- How did you locate the zeroes? (Can you do it another way?)

- What is the value of the y-intercept? (Can you find it another way?)

- If $x = 11$, then what value(s) will y have? (Can you do it another way?)

- If $y = {}^-2$, then what value(s) will x have? (Can you do it another way?)

In each case, responses to the first question provided an assessment of the minimum acceptable performance and responses to the second question encouraged the student to demonstrate an understanding of the process. Students received a copy of their performance assessment at the end of the conference.

The third form of performance assessment employed in *Algebra Through Technology* was a project that involved the design of an experiment. The collection and analysis of data was an integral part of the entire course. Early in the semester, these experiments were carefully controlled in order to address specific algebraic and calculator skills. For example, one assignment involved the analysis of a year's utility information, such as average daily cost of gas or electricity, average monthly temperature, and average daily use of gas or electricity. The question under consideration was how to best predict your monthly utility costs (NCTM, 1995).

As the semester progressed, however, the experiments became more open-ended and required greater understanding of the practical aspects of the situation. For example, for a group project in the final month of the course, students were assigned an experimental situation and asked to produce a well-designed experiment that could be effectively used in a middle school algebra classroom. The original idea, *It's Simply Marbleous,* is from materials produced by the AIMS Education Foundation (AIMS, 1987). Each small group completed the original experiment that involved exploring the effect the slope of a ramp had on the distance a marble would roll. They then decided what other factors might affect the distance and designed a complete experiment which included the original experiment and their extension. The group was expected to devise a set of probing questions and an assessment model for their experiment. The final result was presented to the class in both written and oral form. Their classmates evaluated the experiment informally, making suggestions and asking for clarification, and the instructor completed the formal evaluation.

For preservice teachers, peer evaluation of their experiment was very informative. Viewing the situation from the perspective of a student and a teacher gave each of them simultaneous insights into the difficulties of clear communication in assignments and into the amount of effort required to produce a worthwhile performance assessment task.

Conclusion

An assessment program must provide information about each students' progress toward understanding mathematics. It must also help teachers make instructional decisions. The performance assessments described above accomplished these objectives efficiently and effectively. To paraphrase Gail Burrill (1997), president of the National Council of Teachers of Mathematics, in order to improve mathematics education, you must "show me the mathematics." And in order to evaluate students' progress, they must "show us the mathematics."

References

AIMS Education Foundation. (1987). *Math + science: A solution.* Author.

Ann Arbor Public Schools. (1993). *Alternative assessment.* Palo Alto, CA: Dale Seymour Publications.

Burrill, G. (1997). President's message. *NCTM News Bulletin, 4,* 3.

National Council of Teachers of Mathematics. (1989). *Curriculum and evaluation standards for school mathematics.* Reston, VA: Author.

National Council of Teachers of Mathematics. (1995). *Algebra in a technological world*. Reston, VA: Author.

Stenmark, J. K. (1991). *Mathematics assessment, myths, models, good questions, and practical suggestions*. Reston, VA: National Council of Teachers of Mathematics.

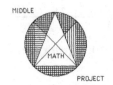
The Effective Use of Technology in Mathematics Teacher Preparation

Zhonghong Jiang and Edwin McClintock
Florida International University

Advanced technology is increasingly pervasive in everyday life. More and more educators believe that the use of technology can effectively facilitate the teaching and learning of mathematics. This belief has been reinforced since the appearance of innovative technologies in mathematics education, including dynamic geometry software such as the *Geometer's Sketchpad* [GSP] (Jackiw, 1991), computer algebra software such as *Maple* and *Mathematica*, spreadsheet programs such as *Microsoft Excel*, graphing calculators such as the TI-83 and TI-92, and a variety of other powerful electronic tools. These technologies are highly interactive so that whenever a student's actions yield a reaction on the part of the machine, it in turn sets the stage for interpretation, reflection, and further action on the part of the student. Using these technologies, one can make powerful resources immediately available to aid thinking or problem solving, provide intelligent feedback or context-sensitive advice, actively link representation systems, and generally influence students' mathematical experience more deeply than ever before (Kaput and Thompson, 1994). In addition, the rapid computing speed of computers and graphing calculators can free students from tedious calculations and allow them to concentrate on conceptual understanding. By opening a new, colorful world to students, technology can greatly motivate students, stimulating their stronger interest in mathematics. Based on these considerations, the NCTM *Standards* (1989) for grades 5-8 emphasize the effective use of technology as one of the chief features of the reform curriculum. In recent years, many research studies (Edwards, 1991; Jiang, 1993; Manouchehri, 1994; Olive, 1991; Thompson, 1992; Thompson and Thompson, 1990) have provided evidence supporting the belief that students benefit from the use of technology.

Teachers are the most important factor in the use of technology in mathematics education. Without qualified teachers who are interested and have enough knowledge, experience and confidence in integrating technology into their mathematics teaching, students could not gain the benefits mentioned above. In order to become qualified teachers, our preservice instructors must receive adequate technology training in our teacher preparation programs and have an appropriate amount of practical experience in using technology with middle school students.

Technology in Teacher Preparation

Our position is that the effective use of technology should be a necessary and important component of our middle school mathematics teacher preparation programs. We should not only integrate the use of technology into all program courses wherever appropriate (especially mathematics content and methods courses) and involve it in students' field experiences, but also have at least one specific technology course which focuses on learning mathematics with technology (i.e., conducting technology-based sense making, problem solving, and mathematical reasoning) and exploring ways of teaching mathematics with technology.

There are several major reasons for us to have this position. First, in order for their future students to become strong mathematical problem solvers in technology-rich environments, the preservice teachers themselves should first be strong problem solvers. Second, they should be familiar with (or at least know the most important features of) different technologies, such as various software packages, graphing calculators, Calculator-based Laboratory (CBL), Multimedia, and Internet so they will be adept at choosing the right technology for best enhancing their future students' mathematical learning in any specific learning situation. Third, most inservice teachers know very little about the up-to-date technology, have little opportunity to consider a richer approach to their teaching, and lack necessary materials and methods for using technology as a teaching tool. Because most districts in the United States offer only approximately four days of release time per year for professional development, the inservice training is, generally speaking, limited. In addition, the one-day or two-day workshop training format can only give teachers a general sense of a specific technology. It is not sufficient to help teachers reach a degree of proficiency in using that technology. It is very important that there are technology experts among the teachers, who can offer long-term and persistent help when needed. In this sense, it is significant for our graduates to become sophisticated in using technology so that they can play the role of such experts (or more accurately, agents of change) after being assigned to schools. While being new teachers who need help with teaching strategies and ways to work with children, our graduates can effectively help inservice teachers with using technology. As a matter of fact, several new teachers who graduated from our program in recent years have been doing a great job as technology experts at the schools where they are working.

Without a systematic preparation in using technology for learning and teaching mathematics, it is not realistic to expect our preservice teachers to handle these tasks well when they take teaching positions at schools. It is also not realistic to provide the systematic technology training simply by using pieces of technology in different courses where instructors must consider more factors than the use of technology.

Carefully designed technology courses can serve the important purpose of substantially strengthening the mathematical background of preservice teachers. Great opportunities are offered for them to revisit school mathematics, to construct deeper understanding of basic mathematical concepts through investigations and explorations, and hence to develop stronger problem solving and mathematical reasoning abilities. This is especially significant for states like Florida, where a legislation-fixed upper limit of 120 hours to graduation has depleted opportunities for enriched teacher training programs and almost no instruction time can be used to offer courses such as *Geometry for Middle School Teachers* and *Algebraic Concepts for Middle School Teachers*.

Two related issues are important to consider when designing a degree program involving a major emphasis on technology. First, different technologies offer different opportunities, so it is inappropriate to favor one technology over another. For example, it is inappropriate to use calculators (even graphing calculators) to the exclusion of computers. Computers offer many major benefits over calculators, including high-resolution graphics and strong animation features. Further, a computer-trained individual is educated in manners consistent with society's needs and the individual's need for changing job requirements and the diversity of occupations that an individual will encounter in his/her lifetime. Calculators are a substantively less useful tool, but have the advantage of being cost effective.

A second important issue is that technology can and should be learned with and in the context of another discipline or area of application. It is necessary for preservice teachers to be productive as members of society—having knowledge and skills in multiple disciplines or multiple application areas. Likewise, in preparing to teach mathematics in an interdisciplinary, connected way, prospective teachers should learn mathematics in an interdisciplinary, connected way.

Technology in Content and Methods Courses

The mathematics education faculty should work collaboratively with mathematicians to explore how to integrate appropriate technology into college mathematics courses. The initial step may include using computer algebra software such as *Maple* and graphing calculators such as the TI-85 or TI-92 in calculus courses. These technologies have been found by a number of mathematicians (Herod, 1996; Kenelly, 1996; Lopez, 1996) to be a great help in facilitating dynamic and interactive visualization for calculus concepts and problems, and in linking the visualization to the numerical and symbolic aspects of calculus to develop students' conceptual understanding. Likewise, we should consider using statistics software, such as *Datadesk,* and calculators with statistics components in statistics and probability courses.

Within the mathematics methods course, technologies should be used for investigations, applications, communications, problem solving, and as a suggested teaching tool. For example, we presented the following problem to our preservice teachers: "A traveler wishes to visit each of five cities in series, beginning and ending in the same city. However, the traveler does not wish to visit the same city more than once. What is the cheapest circuit of the cities that the traveler can make?" Students used various methods to analyze this problem, trying to understand it and solve it. Some of them chose to use technology. Figure 1 shows one of the solutions presented in a spreadsheet.

Leg	Price
Atlanta to Chicago	169
Atlanta to LA	319
Atlanta to Miami	198
Atlanta to NY	279
Chicago to LA	349
Chicago to Miami	329
Chicago to NY	179
LA to Miami	439
LA to NY	419
Miami to NY	249

This is the cheapest route. For reasons of economy, LA to Chicago appears as Chicago to LA. However, if the prices are changed, the route will be updated. Unfortunately, the prices must be kept at least very slightly unique for the text to update properly.

Atlanta to LA
Chicago to LA
Chicago to NY
Miami to NY
Atlanta to Miami

The real route will never-theless be found, and the
Total cost = 1294   total cost will reflect this.

The database below lists all 12 of the possible routes passing through each city once. Since every city is connected to every other, we begin with 5! permutations for visiting the cities. However, the starting city doesn't matter, so by choosing always to begin with Atlanta, we reduce the possibilities to 4!. But the direction of travel doesn't matter either, so there are finally only 12 possible routes.

Leg1	leg 2	leg 3	leg 4	leg 5	Total
169	349	439	249	279	1485
169	349	419	249	198	1384
169	329	439	419	279	1635
169	329	249	419	319	1485
169	179	419	439	198	1404
169	179	249	439	319	1355
319	349	329	249	279	1525
319	349	179	249	198	1294
319	439	329	179	279	1545
319	419	179	329	198	1444
198	329	349	419	279	1574
198	439	349	179	279	1444

Leg1	leg 2	leg 3	leg 4	leg 5	Total
319	349	179	249	198	1294

Price	Price	Price	Price	Price
319	349	179	249	198

These little "databases" were used as extract ranges for the cheapest route, and then to help convert the result into text.

Figure 1. One solution of the circuit problem in the methods class.

Specific Technology Course(s)

If possible, two specific technology courses should be designed and offered to the middle school preservice teachers. One of them could be named *Learning Mathematics with Technology* (treating the preservice teachers as learners of school mathematics in technology-rich environments). The other could be named *Teaching Middle Grades Mathematics with Technology* (concentrating on technology-based teaching ideas and strategies). If, for the time being, it is difficult to offer two technology courses, the availability of at least one such course is a must. In this situation, the course should cover both learning and teaching aspects mentioned above, and be designed around content strands recommended by the NCTM *Standards*. Because the middle grades mathematics that the preservice teachers are expected to teach is the direct foundation for and transition to high school level mathematics, a sound understanding of the two levels of school mathematics is necessary for all preservice teachers. Therefore, the course should deal with both educational levels and balance well between them.

The emphasis of this course should be on exploration of various mathematics contexts to learn mathematics, to pose problems and problem extensions, to solve problems, and to communicate mathematical demonstrations by using various software applications and graphing calculators. Students' classroom explorations, follow-up investigations, and projects that require the full range of mathematical work in technology-rich environments should be major course activities. Sound pedagogical issues such as cooperative learning with technology, journal writing, and more general considerations of communications about, with, and through mathematics should be woven throughout the course. For the teaching with technology component, each student will also be required to construct a technology-intensive learning environment for engaging middle school students in learning a specific topic or unit of mathematics.

The technology activities in this course should involve the following:

- Using the dynamic movement and dynamic measurement features of GSP to construct and investigate basic geometric shapes, explore transformations of geometric figures, determine the relationships embedded in certain geometric phenomena and their real-world applications, and develop proportional reasoning and logic reasoning based on the electronic geometric models. (Figure 2 shows an example in which our preservice teachers used GSP features to do investigations, formulate and test conjectures, generate the construction, and generate the proof and generalizations about the shortest path for a road between two towns on different sides of a river with only two bridges [Jiang & McClintock, in 1997].)

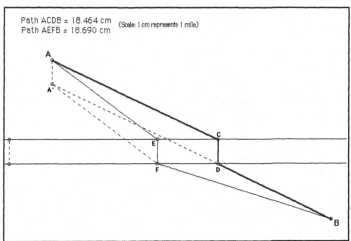

Figure 2. Finding and verifying the shortest path ACDB.

- Using spreadsheets, graphing calculators, CBL, World Wide Web, and other technology tools to explore real-world data collection, data analysis, and statistical concepts through multiple formats such as electronic index cards, tables, charts, and graphs. (Figure 3

shows an example in which our preservice teachers used a spread-
sheet-type statistics program and a learning environment we created
with Hypercard to explore data analysis and statistics concepts—
variance and standard deviation.)

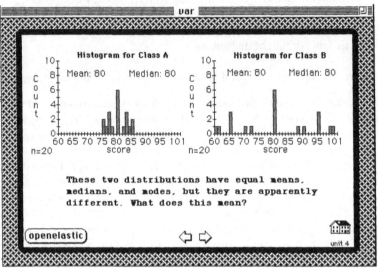

Figure 3. Analyzing test scores to explore measures of variability.

- Using spreadsheets and mathematical microworlds to simulate
  probabilistic phenomena and explore both experimental and theoretical probability. (Figure 4 shows an example in which our preservice teachers played a computerized game via a mathematical
  microworld called CHANCE [Jiang & Potter, 1994] to test their
  conjectures about a probabilistic situation where a *world record* is
  set for the most rolls of two dice without getting doubles.)

- Using GSP, spreadsheets, algebra software, and graphing calculators to explore algebraic operations (within different number systems), patterns and functions with multiple representations such as
  graphical, numerical, and symbolic representations. (Figure 5 shows
  an example in which our preservice teachers used an algebra program called Green Globs and Graphing Equations [Dugdale & Kibbey, 1996] to explore the characteristics of various functions and
  connections between graphical and symbolic representations of
  functions.)

 Special Fast_Process Choice Game Assign_points

Chance World

ROLL DICE   PLAYER1   COPY
PLAYER2   BACK

11 times,
Good record!
Keep rolling!

	Record			Record			Record			Record			Record	
Turn	PL1	PL2	Turn	PL1	PL2	Turn	PL1	PL2	Turn	PL1	PL2	Turn	PL1	PL2
1	1	/	11	/	9									
2	2	/	12	3	/									
3	4	/	13	2	/									
4	8	/	14	0	/									
5	4	/	15	3	/									
6	/	2	16	7	/									
7	/	10	17	/	7									
8	/	4	18	/	5									
9	/	4	19	/	0									
10	/	0												
Rec-ord			Rec-ord	8	10	Rec-ord			Rec-ord			Rec-ord		

Figure 4. Testing probability conjectures by playing *A World Record*.

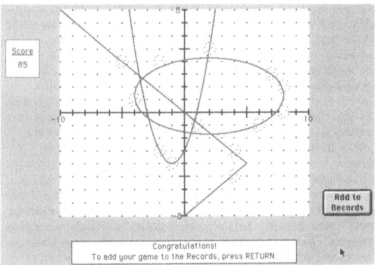

Figure 5. Using Green Globs to explore the characteristics of various
functions and connections between graphs and symbolic equations.
(The three function graphs in the figure were generated by having
entered corresponding symbolic equations.)

Technology in Field Experience

We should require our preservice teachers to implement the appropriate technologies and problem solving techniques learned in the previously described courses with the middle school in their field experience activities. When they go to visit middle grades mathematics classrooms, together with other tasks assigned, they should be asked to study the role of technology in helping develop the children's thinking. It is essential for preservice teachers to have frequent opportunities to observe middle school students' learning with technology. Early in their university training, they must see and experience first hand how middle grades students use technology to support their use of different strategies and thought processes in problem contexts. Prior to student teaching, preservice teachers must have opportunities to introduce small groups of children to learning mathematics with technology. For example, in our teacher preparation program at Florida International University, experiences in working with middle school students are required as part of the fulfillment of both a methods course and a technology course. Examples include using CBL equipment with a TI-82 to collect launch data from a weather balloon launched by the National Hurricane Center on campus, using GSP to explore geometric shapes and transformations in the seventh grade technology class, using spreadsheets, simulation programs, and computer software (e.g., Architech) from the "Middle Math Through Applications Project" with the eighth graders, and using "Computer Intensive Algebra" with technologies such as *Excel* with the ninth graders. These activities provide beginning technology experiences that support students in their study of mathematics and their future academic work. Our preservice teachers participate in these classes, and through class activities are able to involve themselves in ventures that get them into students' thinking in technology-rich environments. For example, they saw the seventh graders become excited about learning mathematics when an animation they had created "almost" produced an elliptical orbit of the earth about the sun and, simultaneously, an orbit of the moon about the earth. Technology with field experience should be available throughout a teacher's preservice training. During their student teaching, it should be required that they use at least one appropriate technology to lead their students to explore related mathematical concepts and problem solving. It is necessary, we believe, for preservice teachers to experience this sort of environment during their junior and senior years in order to be prepared to teach mathematics effectively today (and possibly be able to adopt the changes appropriate to individual and societal needs of tomorrow).

Research on Using Technology

We should conduct research on how our preservice teachers adapt to and use technology in their learning and teaching of mathematics and how this use of technology effects their own and their students' learning and understanding of mathematics, leading to a better understanding of the change our preservice teachers experience and its outcome upon children's learning. A first stage of the research might be to study how preservice teachers originally view the use of technology in the mathematics classroom and their possible attitude change during the course. At the heart of any educational process is an associated change process, and this is particularly acute in a major reform. Some sense of lasting belief systems, values and level of commitment are reflected in the way, the extent and the persistence with which technology is used.

Conclusion

"It has become increasingly evident that the technology altered the nature of the activity using it" (Kaput and Thompson, 1994, p. 677). There is no doubt that technology, if used effectively, can facilitate mathematics education. If we want to design a strong middle school mathematics teacher preparation program that is consistent with the reform of mathematics, we must emphasize the infusion of technology into the program. Important aspects of doing so include integrating the use of technology into mathematics content and methods courses, designing and offering at least one specific technology course, and implementing technology in students' field experience activities. We should constantly improve these practices by conducting research on the ways technology effects teacher preparation, helping us decide which direction to take.

References

Dugdale, S & Kibbey, D. (1996). Green Globs and Graphing Equations [Computer software]. Pleasantville, NY: Sunburst Communications.

Edwards, L. D. (1991). Children's learning in a computer microworld for transformation geometry. *Journal for Research in Mathematics Education, 22*, 122-137.

Herod, J. (1996). *Looking for problems that deserve the power of Maple*. Workshop given at Florida International University, Miami, FL.

Jackiw, N. (1991). The Geometer's Sketchpad [Computer software]. Berkeley, CA: Key Curriculum.

Jiang, Z. (1993). *Students' learning of introductory probability in a mathematics microworld*. Unpublished doctoral dissertation, The University of Georgia.

Jiang, Z., & Potter, W. (1994). A computer microworld to introduce students to probability. *The Journal of Computers in Mathematics and Science Teaching, 13*(2), 197-222.

Jiang, Z., & McClintock, E. (1997). Using *The Geometer's Sketchpad* with preservice teachers. In J. King & D. Schattschneider (Eds.), *Geometry turned on: Dynamic software in learning, teaching and research*. Washington, DC: Mathematical Association of America.

Kaput, J. J., & Thompson, P. W. (1994). Technology in mathematics educational research: The first 25 years in the JRME. *Journal for Research in Mathematics Education, 25*, 676-684.

Kenelly, J. (1996). *New technologies and their effects on the AP Calculus exam*. Workshop given at Florida International University, Miami, FL.

Lopez, R. (1996). *Using Maple in teaching mathematics*. Workshop given at Florida International University, Miami, FL.

Manouchehri, A. (1994). *Computer-based explorations and mathematical thinking processes of preservice elementary teachers: Two case studies*. Unpublished doctoral dissertation, The University of Georgia.

National Council of Teachers of Mathematics (NCTM). (1989). *Curriculum and evaluation standards for school mathematics*. Reston, VA: Author.

Olive, J. (1991). Logo programming and geometric understanding: An in-depth study. *Journal for Research in Mathematics Education, 22*, 90-111.

Thompson, P. W. (1992). Notations, conventions, and constraints: Contributions to effective uses of concrete materials in elementary mathematics. *Journal for Research in Mathematics Education, 23*, 123-147.

Thompson, P. W., & Thompson, A. G. (1990). Salient aspects of experience with concrete manipulatives. In G. Booker, P. Cobb, & T. Mendicuti (Eds.), *Proceedings of the 14th International Conference for the Psychology of Mathematics Education*. Oaxtepec, Mexico.

Examining My Assumptions

Jennifer Earles Szydlik
University of Wisconsin-Oshkosh

Every once in a while I need to be reminded of what it is that I don't know. I am an enthusiastic supporter of the current reform in mathematics education — I love the *NCTM Standards* and my teaching is generally "transformed." I know students construct their own knowledge. I know they learn mathematics best when they are actively solving problems. I know teachers teach as they were taught. I know technology enhances students' learning. Or do I <u>know</u> these things? Is there sufficient evidence to support these claims?

As part of a grant proposal, a colleague and I wanted to make an argument that the tenets of the reform (e.g., those stated above) are true, and that we should receive money to align our course in probability and statistics for preservice middle school teachers with those tenets. However, our literature search revealed less than convincing evidence. For example, we could not find research that suggested students learn probability or statistics better under a "reformed approach." Nor could we argue, based on the literature, that the calculator enhances student learning (there is not a clear consensus among those studies).

In her address at the Second MIDDLE MATH Conference, Joan Ferrini-Mundy reminded me I need to consider the possibility that I don't know how students learn and how teaching is best accomplished. This does not mean the *Standards* are "wrong" or that our work is fruitless. It does suggest that we need to keep exploring, listening, thinking, and studying our assumptions. And it means that we <u>might</u> be wrong, and that we might learn something.

Printed in the United States
By Bookmasters